Water Thievery

WATER THIEVERY

THE ART OF WATER SUPPLY OPERATIONS

ANDY SOCCODATO

FIRE ENGINEERING™
· BOOKS ·

Disclaimer

The recommendations, advice, descriptions, and methods in this book are presented solely for educational purposes. Photos are for instructional purposes only. Always wear the proper level of approved PPE when conducting training drills and operating at incidents. The author and publisher assume no liability whatsoever for any loss or damage that results from the use of any of the material in this book. Use of the material in this book is solely at the risk of the user.

Fire Engineering Books & Videos
110 S. Hartford Ave., Suite 200
Tulsa, Oklahoma 74120 USA
800.752.9764
+1.918.831.9421
info@fireengineeringbooks.com
www.FireEngineeringBooks.com

Executive Vice President: Eric Schlett
Vice President, Group Publishing: Amanda Champion
Acquisitions: David Rhodes, Diane Rothschild
Director, eLearning and Books: Starlet Franz
Production Manager: Tony Quinn
Book Designer: Robert Kern, TIPS Publishing Services, Carrboro, NC
Cover Designer: Brandon Ash

Library of Congress Cataloging-in-Publication Data Available on Request

ISBN: 9781593705947

Printed in the United States of America

2 3 4 5 6 27 26 25 24

This book is dedicated to the loving memory of my mother, Madeline. Thank you for teaching me how to be the instructor I am today.

CONTENTS

FOREWORD

Every once in a while, people just get it right. They choose the right dress for a formal night out, they choose the right flight to avoid delays, or in this case, they choose the right book to not only entertain, but to educate as well. Today is that day for *you.*

There are so many names for the pump operator: *driver, chauffeur, engineer, wagon driver, fire apparatus operator, heavy equipment operator,* and the list goes on. No matter the rank or title, the good ones are proud of their role, take the job seriously, and know that the best pump operator can quickly troubleshoot, think three steps ahead, use muscle memory, and is not afraid to push the limits of the engine, ladder truck, or water system when necessary. This book will provide you with the knowledge you need to take your understanding of hydraulic principles to the next level.

The pump operator has an obligation to perfect their craft. This individual has the sole responsibility to manage the hose streams and water supply, and in many cases in short-staffed fire departments, complete critical job tasks away from the pump panel as well. Remember these principals as you read, and reread, the text before you.

The question is, "Who am I, and what the heck do I know about this?" Let's go back to 2015 when a friend landed the job as the fire chief in a city about an hour away. Despite his moving, we kept in touch, and in a passing conversation, he mentioned that he had a guy in his department that I needed to meet. He recognized our love of pump operations, hydraulics, and class instruction, and he said we might be a good fit together. I remembered that the guy's name was "Andy Something-Weird" and tucked it away in the back of my brain. Fast forward to a chilly February day, when I'm in a class to get some instructor updates, and I see that guy's name on the roster. I introduced myself, and it was the beginning of what has been a great friendship.

Andy Soccodato is a friend, confidant, brother, and uniquely gifted fire service instructor. His brain operates at a different level than most people's. As a self-professed "dork" when it comes to pumping, he has a passion for moving water, performing calculations, lowering friction loss, increasing flow, and making other firefighters as excited about these things as he is.

As Andy and I travel across the United States to deliver "Water Thieves" classes, we each gravitate toward delivering different portions of the practical scenarios. However, I generally only participate in the lecture portion of the class when he asks. Why? Because I find myself sitting in the back of the room taking notes, as I continue to learn more each time he stands in the front of the room engaging the students. It does not matter if it is in a packed room at the Fire Department Instructors Conference or in the apparatus bay at the local volunteer firehouse. Regardless of the location, you will find that his passion for making his students better pump operators is contagious!

Each time I provided Andy with editing advice for his book, my excitement built for him and the great things he is doing for the fire service. I know that when you are done with it, your skills and critical thinking will have been taken to the next step as well.

Be smart, be safe, and be faithful to the job.

—Taylor Goodman, MO
Deputy Chief, Powhatan County Fire Rescue Department

PREFACE

After joining the fire service in 2006, I was immediately drawn toward the role of the driver/operator. This attraction inspired me to immerse myself in the world of pump operations and water supply. During this journey, I have been fortunate enough to have some amazing mentors who further cultivated my love for pump operations by teaching me how to competently serve as a pump operator.

The information presented in this book is the culmination of institutional knowledge passed down to me, real-world experiences, and lessons learned from mistakes that I have made. The intention is to provide the readers with material that will help them better understand water supply operations and the importance of the supply pump operator, regardless of whether the water supply operation involves fire hydrants, rural water shuttles, or a mixture of both. Readers should then take this information and train on it with their own apparatus and equipment to determine what works best for them.

INTRODUCTION

Water supply is arguably one of the most critical tasks that must be completed on the fireground. It does not matter if the fire occurs in the most urban city in the United States or on a rural country road—a water supply must be established in order for any type of aggressive firefighting operations to take place. Water supply operations involve more than just getting water to the attack engine. Effective operations will not only provide the scene with the needed volume of water for the current operation, but they will also be built to support the flow requirements for escalating incidents. This requires the operator of the supply pumper to think proactively.

This text requires the reader to have a basic understanding of pump operations. After reading this text, the reader will have a thorough understanding of more advanced water supply tactics. The principles listed below are the foundation that enable complex water supply operations to take place:

1. *Functionality above all else.* Operators should remember that functional water supply operations are always the most efficient operations. This extends from water supply tactics all the way to apparatus design and layout. By keeping functionality at the forefront, water supply operations will be able to be rapidly established but will also be easy to expand when needed.
2. *Size matters regarding water supply operations.* In hydraulics, area has the biggest impact on the water availability for an operation. Increasing the area on the intake side of the pump involves heavy hydrant hookups and multiple intake drafting operations. On the discharge side of the pump, this takes the form of large discharge plumbing and parallel supply line operations.

3. *Drafting is the most fundamental pump operations skill.* The art of drafting is not just reserved for rural pump operators. It is a skill that should be maintained by those operating in an urban setting as well. Not only is it possible for urban fire departments to be called into a rural setting, but it is also possible to implement rural water supply tactics in a city environment. This could be necessary due to a failure of the municipal water distribution system or in other situations in which a supplemental water source is needed. In any event, it will be critically important that the operators involved with the water supply operation are proficient with their drafting skills.
4. *Build the water supply operation in segments.* In order to remain functional, water supply operations must be built in segments. One of the common mistakes regarding water supply operations is that firefighters attempt to go too big right from the beginning. Instead, the configuration should start simple and build complexity as equipment, time, and resources become available. Remember, we do not go from "zero to octopus." Instead, we build our system one "tentacle" at a time.
5. *Get out and train on water supply operations.* It is quite easy to put off any type of training that involves water supply, whether it is because it may cause apparatus to be out of service or because firefighters simply do not want to load hose. Putting off this vital training puts the operator at a severe disadvantage, however. Take the information presented in this text and utilize it for your water supply training. Remember, training on water supply does not always have to involve multicompany scenarios flowing several thousand gallons per minute. Water supply discussions around the kitchen table are some of the best sources of training on this topic.

1

COMPONENTS OF MODERN PUMPING APPARATUS

The introduction of the automotive fire apparatus revolutionized the fire service in the early part of the 20th century. For more than 100 years, the automotive fire apparatus has been the workhorse of the fire service. This tried-and-true design concept has gone through several major improvements that have increased firefighter safety, operational effectiveness, and the volume of water that can be delivered for firefighting operations (fig. 1–1).

Figure 1–1. Automotive fire apparatus with centrifugal fire pumps are the workhorses of the modern fire service.

It is critically important for pump operators to be intimately familiar with every component on their apparatus. Such knowledge will enable them to operate effectively, troubleshoot problems rapidly, and reliably supply water to the fire scene. This chapter will refresh the baseline knowledge of pump operators regarding pumping apparatus components while introducing key principles that will enable them to perform more complex water supply operations.

Centrifugal Fire Pumps

The *centrifugal fire pump* is undoubtedly the workhorse of the modern fire service, and it is used as the main fire pump on apparatus manufactured today. The centrifugal fire pump is defined as a non-positive-displacement pump, which means two critical things to the pump operator:

1. The centrifugal fire pump does not move a definitive amount of water with each rotation.
2. The centrifugal pump cannot pump air; it can only pump water.

Since centrifugal fire pumps cannot pump air, they are not self-priming. To overcome this problem, positive displacement pumps are installed on modern fire apparatus to serve as priming pumps for the main centrifugal fire pump. Priming pumps are required in situations where the centrifugal fire pump is completely drained of water. If the centrifugal pump is full of water, it is possible to perform alternative drafting techniques that will enable a prime to be achieved from an outside source without the use of a primer pump. These methods will be discussed in chapter 4.

The centrifugal pump works using centrifugal force to move water. These pumps impart velocity to the water and convert it into pressure within the pump. Initially this may seem like a confusing concept to grasp. When we break it down by looking at how each component within the pump operates, however, a clear picture emerges.

In these pumps, there is a rapidly revolving disk, called the *impeller*, mounted on a drive shaft and housed within a pump casing. Water from a source is directed through intake plumbing toward the *eye* of the impeller (the center of the disk). The water enters the eye of the rapidly revolving impeller, where it is directed toward the outer edge by the vanes inside the impeller itself. While the impeller is spinning, it throws the water outward via centrifugal force.

At this point, the spinning impeller has utilized centrifugal force to impart velocity to the water being thrown outward. The faster the impeller spins, the more velocity is imparted to the water. The water that is discharged from the impeller is sent into the *volute*, which is a portion of the pump casing that increases gradually in cross-sectional area. This critical component of the centrifugal pump allows the velocity imparted on the water to be converted into pressure.

These components must come together in order for this process to work. Without a casing and volute, the spinning impeller would be nothing more than a giant sprinkler. Without the impeller, the casing and volute would be a useless chunk of metal. When the pump is full of water and the impeller is spinning, the water is first sent to the volute, where pressure is built, and then subsequently to the discharge manifold of the pump. The discharge of water from the pump naturally creates an area of negative pressure (a *vacuum*) at the eye of the pump's impeller. This means that water will continue to flow as long as the following conditions are met:[1]

1. No significant amount of air enters the intake side of the pump.
2. The supply of water entering the pump, either from a draft or hydrant, is sufficient to meet the discharge demands (i.e., the amount of water being discharged is not greater than the amount of water coming in).
3. While drafting, the loss due to lift is not greater than the friction loss in the hard sleeve for a given flow.
4. There are no obstructions in the intake strainer(s).

For most municipal applications, centrifugal fire pumps are configured as either *single-stage* or *two-stage pumps*. Single-stage centrifugal fire pumps utilize one impeller mounted on an impeller shaft inside the pump casing. Two-stage pumps have identical impellers mounted on the same impeller shaft. The advantage of the two-stage pump is that it has the ability to pump in a series position, pumping water from the first impeller directly into the second impeller. This allows higher pump pressures to be developed at lower engine speeds or revolutions per minute (rpms). The trade-off to operating in this position is that the pump can only be expected to deliver 50%–70% of its rated capacity. Most U.S. fire departments currently utilize single-stage pumps.

Pump operators must understand that a two-stage pump differs from a *pressure governor.* The volume position and pressure position of a two-stage pump often get confused with the *pressure mode* and *rpm mode* of a pressure governor. Readers must remember that a pressure governor is an automatic pressure control device that helps regulate discharge pressures. A two-stage pump is a type of pump design that utilizes two impellers to either deliver its volumetric capacity or to work in a series fashion to build high pressures at low engine revolutions per minute. Pressure governors will be discussed in further detail later in this chapter.

Pump capacities of centrifugal fire pumps

When operating a centrifugal fire pump, it is critically important to understand what is signified by the *pumping capacity* of that pump. In order to understand the pump rating of a particular apparatus fire pump, the operator must first understand the difference between *volume* and *pressure. Volume* is defined as the amount of water flow over a defined period of time. In the U.S. fire service, this is expressed as gallons per minute (gpm). *Pressure* is defined as force per unit of area. In this text, pressure will be expressed in pounds per square inch (psi). Three major variables dictate the rated pump capacity or volume that a centrifugal fire pump can achieve:

1. The size of the pump's impeller
2. The number of pump discharges
3. The horsepower of the vehicle's motor

These components come together to allow the centrifugal pump to move its rated volumetric pump capacity. *National Fire Protection Association (NFPA) 1900: Standard for Aircraft Rescue and Firefighting Vehicles, Automotive Fire Apparatus, Wildland Fire Apparatus, and Automotive Ambulances* requires that municipal pumpers have a minimum pump capacity of 750 gpm, although pump sizes of 2,000 gpm are becoming more popular in the U.S. fire service.[2] These pump ratings increase in increments of 250 gpm. Most pump manufacturers use the same impeller size and casing for pumps rated between 750 gpm and 1,250 gpm. For pumps rated for 1,500 gpm to 2,000 gpm, the impeller and casing sizes are also the same. This means that the major difference between a 1,500 gpm pump and a 2,000 gpm pump will be that the 2,000 gpm pump will have more discharges and may have a larger horsepower motor.

NFPA 1900 also states that the pumper must have enough 2½" or larger discharges installed to be able to flow the pump's rated capacity.[3] The plumbing size specified by the department when the apparatus is built will determine the minimum flow rate through the discharges and ultimately the total number of discharges that will be on the apparatus. The biggest takeaway here is that apparatus with larger pump capacities will be equipped with more discharges in order to supply the rated capacity.

The final variable that affects a centrifugal fire pump's rated pump capacity is the horsepower of the vehicle's motor. Moving large volumes of water requires the impeller to spin extremely fast. In order to do this, the apparatus motor must be able to produce enough torque to spin the impeller fast enough to discharge a high volume of water. It is important to note, however, that larger horsepower motors also come with much higher price tags. Many fire departments balk at the idea of potentially spending an extra $30,000 to $50,000 for a larger motor.

Apparatus manufacturers may suggest that a smaller 450 horsepower (HP) motor can power a 2,000 gpm pump and save a large up-front cost. While this is true, the purchaser must remember that the smaller horsepower motor may be able to pump 2,000 gpm, but it will be the upper limit of the vehicle's capacity. This will show itself every year during pump testing. The larger 600 HP motor may be more expensive initially, but it will move its capacity with ease for the duration of its service life (fig. 1–2).

There are three variables that influence the amount of pressure that a centrifugal fire pump is able to produce:[4]

1. The pressure of the water when it enters from a positive pressure source
2. The speed of the impeller if volume remains constant
3. The volume of water being discharged from the pump

One of the major advantages of the centrifugal fire pump is its ability to make use of existing pressure. For example, if we were to hook our pumper to a fire hydrant and turn the hydrant on, the pressurized water coming into the pump would register a reading on the *master intake gauge*. Let us say the intake reading is 50 psi from the hydrant. If you had to pump an attack line at 150 psi, the pump would only have to add 100 psi to the 50 psi already coming into the pump. The hydrant pressure essentially helps the pump build the required pressure, and thus the pump and the apparatus

Figure 1–2. Larger horsepower motors are extremely beneficial during heavy flow operations.

motor do not have to work as hard to produce the same result. It is for this reason that centrifugal fire pumps are rated while drafting. During a drafting operation, the centrifugal pump receives no help. It not only has to build all of the required discharge pressure, but it also has to create a vacuum to draw water into the pump.

Another way that pressure can be built within a centrifugal pump is to increase the speed at which the impeller is spinning. When the flow rate leaving the pump remains the same and the impeller speed is doubled, the pressure within the pump will quadruple.[5] What is critical to remember here is that the flow rate leaving the pump must remain the same in order for a proportional increase in pressure to occur.

The final variable that influences the pressure produced by a centrifugal pump is the volume of water being discharged from the pump. This last variable is the one that typically tends to confuse firefighters. Common sense tells us that if we want more water, we need to build more pressure or throttle up. Centrifugal fire pumps are designed to flow higher volumes of water at lower pump pressures, however. If a higher pressure is required, the flow rate has to be reduced in order to build the desired pressure.

The easiest way to visualize this is to picture a garden hose without a nozzle attached. With the garden hose flowing, we could expect to fill a five-gallon bucket pretty quickly. The bucket would have to be quite close to us, however, since the flow would not have much reach. If we wanted the stream to go a further distance, we could simply place our thumb over the open end of the garden hose. The reduction in the discharge area will reduce the flow rate and cause pressure to be built. This will allow the stream to reach a greater distance.

Centrifugal fire pump ratings become extremely clear once we examine the variables that affect pressure and volume. When we look at a pump certification plate, we see that centrifugal fire pumps are rated to flow 100% of their pump capacity at 150 psi, 70% of their capacity at 200 psi, and 50% of their capacity at 250 psi while drafting (fig. 1–3). In order to build pressure to 200 psi or 250 psi, the discharge flow rate must be reduced. This is accomplished by gating back or closing discharges until the pump is delivering either 70% or 50% of its capacity, depending on which test is being performed. If the discharge valves were not gated back, the operator who simply tried to throttle up to 200 psi or 250 psi would experience an increase in the flow rate with little increase in discharge pressure. This is because the discharge opening is too large, and the water leaves the pump too rapidly for pressure to be built. By restricting the discharge opening, the volumetric flow rate is reduced, and pressure can be built.

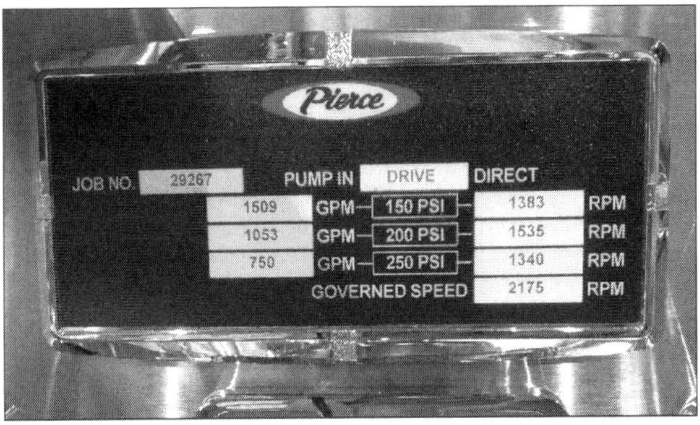

Figure 1–3. The pump certification plate on a 1,500 gpm pumper

The operator must remember that centrifugal fire pumps are only designed to do one of two things at any given time: they can either flow high volumes of water at low pump pressures or flow low volumes of water at high pump pressures. This concept will become critically important in later chapters with regard to large flow operations. In the following sections, we will examine the critical components that make up the intake and discharge sides of the pump. Having a solid understanding of these components will allow the operator to master more complex water supply operations.

The Intake Side of the Pump

The centrifugal fire pump is an incredible piece of machinery that enables the fire service to be successful on the fireground. In order to work effectively, water has to enter the eye of the pump's impeller from a water source. In this section, we will examine the major components found on the intake side of the centrifugal fire pump. These include the tank-to-pump valve and plumbing, pump intakes, intake valves, intake relief valves, and the master intake gauge.

The tank-to-pump valve and plumbing

Centrifugal fire pumps receive their water supply from only one of three potential sources:

1. The onboard apparatus water tank
2. An external pressurized water source
3. An external static water source

The initial water utilized for the vast majority of fires typically comes from the onboard apparatus water tank or booster tank. The tank-to-pump plumbing and tank-to-pump valve enable the water carried in the onboard booster tank to enter the eye of the impeller and be utilized for initial firefighting operations. *NFPA 1900* specifies what the tank-to-pump plumbing and valve arrangement must be capable of flowing based on the tank capacity of the booster tank. If the booster tank is less than 500 gallons, the tank-to-pump plumbing and valve must be capable of flowing at least 250 gpm to the pump. Apparatus equipped with booster tanks 500 gallons or greater must have tank-to-pump plumbing and valve arrangements that allow at least 500 gpm to be delivered to the pump.[6]

The tank-to-pump flow rates mentioned above are considered the baseline minimum requirements for pumping apparatus. Unless a fire department requires and specifies a higher tank-to-pump flow rate when they are ordering their rig, most manufacturers will design the plumbing to meet these minimum requirements (figs. 1–4 and 1–5). It is critically important for pump operators to make a mental note of their tank-to-pump flow rate because it will directly affect them during attack and supply engine operations.

Figure 1–4. Fire departments that desire to flow more than the *NFPA 1900* minimum 500 gpm from their tank-to-pump plumbing may specify that their apparatus be outfitted with dual tank-to-pump valves. Note on this older rig that both are labeled "Tank Discharge Valve."

Figure 1–5. This pumper is outfitted with two tank-to-pump valves and is capable of flowing 1,000 gpm from its booster tank.

Pump intakes

The pump intakes allow water from external sources to enter the pump. Intakes come in a variety of configurations based on the desires and needs of a particular fire department. Common intakes found on modern pumping apparatus include *side intakes, front intakes, rear intakes,* and *auxiliary intakes.* According to *NFPA 1900,* there must be a sufficient number and size of intakes for the pump to meet its certification rating.[7] Most midship pumpers manufactured and rated between 1,000 gpm and 2,000 gpm come with a minimum of two 6" pump intakes. In order to perform more advanced pumping operations, however, it is recommended that the apparatus have at least one intake on all four sides of the rig.

Side intakes are considered to be the best intakes on a pumping apparatus because they are typically plumbed straight into the eye of the impeller. On most municipal pumpers, these intakes are plumbed with 6" pipes leading to the impeller. This combination allows large flow rates to move through the side intakes with a minimal amount of internal loss. During large flow operations, it is always recommended to utilize a side intake as one of the intakes receiving water from an external source.

Front intakes are intake connections located on the front bumper of the apparatus that feed water to the pump (fig. 1–6). These intakes are

Figure 1–6. Front intakes provide the pump operator with unparalleled positioning advantages.

extremely useful from a positioning standpoint, but their flow characteristics typically differ from those of the side intakes. In order to move water from the front intake to the impeller, the plumbing for these intakes must be routed over the front axle of the vehicle in some fashion. This is accomplished by utilizing elbows in the plumbing to get water up and over the front axle. These elbows and bends in the front intake plumbing result in an internal loss of flow rate due to friction and therefore reduce the flow rate to the pump. It is also important to note that the actual internal plumbing that runs to the pump may only have an internal diameter of 5" or smaller.

These characteristics of front intakes cause many firefighters to believe they are no longer useful in the fire service. While these intakes do not allow for the same flow rates as side intakes, they do offer pump operators more flexibility with regard to positioning and the initial establishment of a water supply. This allows for greater functionality in the water supply operation in the long term. Throughout this text, the utilization of front intakes for the fast initial establishment of a water supply will be discussed. In addition, options for expanding the water supply operation to allow for higher flow rates will also be discussed. Pump operators must always remember that functionality is the key to success with regard to any water supply operation.

Similar to front intakes, rear intakes allow a supply line to be connected to the rear of the apparatus to supply water to the pump. The plumbing configurations of rear intakes will vary depending on the pump and apparatus manufacturer. Some manufacturers plumb their rear intakes in a similar fashion to a front intake and incorporate multiple elbows. This obviously causes the same restriction in flow as discussed with front intakes. Some manufacturers plumb their rear intakes extremely efficiently, however, and the result is flow capabilities that rival those of side intakes (fig. 1–7). For these reasons, it is critically important for pump operators to be familiar with their apparatus and to understand the plumbing configurations for each of the intakes on the rig.

Auxiliary intakes refer to the smaller 2½" intake connections located on the pump panel. These connections are ideal for booster back up or direct pumping operations. In addition, these intakes can be used during *heavy hydrant hookups* to assist with gaining additional water from a hydrant. Each of these operations will be discussed in later chapters.

Figure 1–7. The rear intake of this pumper is capable of flowing big water due to the plumbing arrangement inside the apparatus.

Intake valves

Intake valves serve a critical purpose on modern pumping apparatus because they allow for the control of water into the pump. *NFPA 1900* requires that at least one intake on the apparatus be valved and controlled from the operator's pump panel.[8] Technically speaking, the quarter-turn valve on a 2½" auxiliary intake meets the intention of the standard. This means that you could purchase a 2,000 gpm pumper with two 6" side intakes and one 2½" auxiliary intake and only have a valve on the auxiliary intake. In this scenario, the two 6" side intakes would have blind caps attached. While this meets the letter of the standard, it does not make this pumper very functional on the fireground. Although it is not required by *NFPA 1900*, it is highly recommended that every intake on the apparatus be equipped with an intake valve. This is primarily because it gives pump operators options when hooking up their water supply.

Intake valves come in a variety of shapes and configurations, but they are typically characterized as either *internal intake valves* or *external intake valves*. Internal intake valves are beneficial because they do not typically cause the intake connection to extend past the side of the apparatus. In addition, most internal intake valves have larger waterway openings than their external counterparts. Some of these valves currently manufactured have internal waterways of up to 6⅖".[9] This results in higher flow rates through the intake valve and does not cause a reduction in pump capacity during apparatus pump testing. The only major disadvantage of internal intake valves is that when repairs to the valve must be made, it often requires that the fittings, the panels, and potentially even the plumbing be disassembled and removed to reach the internal intake.

External intake valves are a common option for controlling the flow of water into the centrifugal fire pump. These valves simply thread onto the intake plumbing of the apparatus (fig. 1–8). The major advantage is that they can be installed on the apparatus at any point during the service life of the vehicle. These valves can also be removed and repaired relatively easily in the firehouse. The major disadvantage of the external intake valve is that depending on the brand or style of valve utilized, the waterway opening can be extremely choked down. There are external intake valves currently on the market that only have a 3½" internal waterway. The reduction in waterway size will have a significant impact on the volume of water that is able to enter the pump. For this reason, many external intake valves must be removed from the pumper during apparatus pump testing in order for the pump to reach its rated capacity.

Operators should remember that not all external intake valves are bad. There are manufacturers today that produce external intake valves with larger waterways to allow higher volumes into the pump. One valve currently on the market is equipped with a 5¼" internal waterway. This valve equipped with a 5" Storz intake connection can flow 2,000 gpm with only

Figure 1–8. The external intake valves on this quint allow the flow of water to be controlled without sacrificing the flow rate. The waterway on these valves is 5¼".

5 psi of internal pressure loss. If the fire department specifies the valve with 6" male threads for drafting operations, this valve is capable of flowing 2,000 gpm with only 3 psi of internal pressure loss.[10]

Bleeder valves are another key component of intake valves. The intention of a bleeder valve is to exhaust air from the supply line prior to opening the pump intake. This prevents air from entering the pump and causing a loss of prime, erratic nozzle reactions, and injury to firefighters. *NFPA 1900* requires that every valved intake on the apparatus be equipped with a bleeder valve.[11] For internal intake valves, the bleeder valve controls are usually located on the pump panel. External intake valves typically have a bleeder valve located directly on the valve itself.

Intake valves, regardless of whether they are internal or external, should be placed on every apparatus intake as a best practice. This not only gives pump operators more options when positioning as to where they can access their feed, but it also allows superior flow rates to be achieved. Simply put, when pump operators can control the flow of water entering the pump, they can control how much water leaves the pump. Throughout this text, the benefits of intake valves during advanced water supply operations will become quite evident.

Intake relief valves

Intake relief valves are designed to prevent excessive pressures from entering the pump and are required according to *NFPA 1900*. The standard requires that all valved intakes 3" or larger have intake relief valves on the supply side of the valve. This helps protect the supply hose from bursting when a supply line is charged and pressure is built, but the intake valve has not yet been opened. If the apparatus is designed with no intake valves on the large pump intakes (3" or larger), then the standard requires that an intake relief valve be installed on the intake plumbing of the pump to prevent excessive intake pressures.[12] These will be referred to as *internal intake relief valves* in this text.

Intake relief valves must discharge to the open atmosphere and must discharge away from the pump operator.[13] On rigs with internal intake relief valves, the discharge pipe is usually under the pump panel. The relief valve discharge will be equipped with 2½" male threads and typically will have a label that says "DO NOT CAP" somewhere on the piping. The reason for these threads is so that water can be redirected from the intake relief valve using a 2½" hose during operations in freezing conditions or if the rig is on an unpaved road or driveway.

If the intake relief valve discharge is capped, it renders the intake relief valve useless. Another concern occurs if trash gets in the intake relief valve since it will prevent the valve from seating all the way. This will let air enter the intake side of the pump, causing either a loss of prime or the inability to achieve a prime during a drafting operation. If the pump operator cannot achieve a prime after verifying that the connections are tight and drains are closed, one troubleshooting option is to cap the intake relief valve discharge to see if that is the culprit. This is the *only* time this option is acceptable because there is no chance of excessive positive pressures on the intake side of the pump when the pumper is drafting. After the drafting operation is completed, the blind cap *must be removed* from the discharge of the intake relief valve.

The master intake/compound gauge

One of the most important gauges on the pump panel is the *master intake gauge*. This gauge is considered to be a *compound gauge* because it can read both positive and negative intake pressures.[14] *NFPA 1900* requires the master intake gauge to have a scale that reads from 30" of mercury (inHg) to at least 300 psi.[15] The master intake gauge measures the pressure of the water just before it enters the eye of the impeller. When receiving water from a pressurized source, the master intake gauge will read on the positive side of the gauge's scale. During a drafting operation, the master intake gauge will read in the vacuum portion of the scale.

The information provided on the master intake gauge is invaluable to the pump operator, and it is critically important for the pump operator to constantly monitor this gauge. Most operators are vigilant about watching their master intake gauge when receiving a positive pressure water supply. During drafting operations, however, many engineers tend to be less attentive. This is primarily because the vacuum portion of the scale of many compound gauges comprises only a small percentage of the total gauge scale.[16] Since the vacuum scale is so small on the traditional gauges used, it results in very little movement of the intake gauge, and many operators tend to ignore the information the gauge is providing (fig. 1–9). The information provided by the master intake gauge during hydrant and drafting operations will be examined in more detail in chapters 4 and 7.

Figure 1–9. The master intake/compound gauge of a fire department pumper. Note how small the vacuum scale is on this gauge. This pumper is currently drafting and is showing a vacuum reading of 15 inHg.

The Discharge Side of the Pump

After water travels through the intake plumbing leading to the pump's impeller, it is discharged into the volute and to the discharge manifold. From here, the water can be sent in a variety of directions. In this section, we will examine the components that make up the discharge side of the pump and the mechanisms in place to help prevent damage or injury due to excessive pressure surges.

The tank fill valve and plumbing

The *tank fill* is the most basic discharge on the entire pump. This line simply runs from the discharge side of the pump and sends water into the booster tank. Its primary purpose is to refill the apparatus booster tank after a water supply has been established. It is also commonly used to recirculate water during pumping operations to help keep the pump from overheating or losing a prime during drafting operations.

NFPA 1900 lists specific requirements relating to the tank fill valve and plumbing. According to the standard, if the apparatus is equipped with a booster tank of less than 1,000 gallons, the tank fill line must be at least 1" in diameter. For tanks that are 1,000 gallons or greater, the tank fill line must be at least 2" in diameter.[17] The pump operator should consider the tank fill line a self-contained discharge within the pump.

The size of the tank fill line and how far the tank fill valve is opened will have a direct impact on the volume of water flowing into the booster tank. The further the valve is opened, the faster the tank will fill, but this could have negative impacts as well, especially on apparatus with larger tanks and fill lines. When operating solely from the apparatus water tank, if a rig equipped with a 2" tank fill line has the tank fill valve fully opened, the pump will likely cavitate when the throttle is increased. This is because the fill line can flow more water than the tank-to-pump line can supply. It is why in many instances apparatus with larger water tanks, and therefore larger tank fill lines, may only be able to partially open their tank fill valve when operating from their booster tank.

The pump operator must also be cognizant of how far the tank fill valve has been opened during large flow operations. Since the fill valve is considered a discharge, it takes a portion of the water coming into the pump. During a large flow operation, it is not necessary to recirculate water to keep the pump cool or to maintain a prime. Supplying heavy-caliber streams will naturally accomplish both of these goals. If the operator leaves the tank fill valve partially or completely open, it will rob water from the firefighting operation and simply dump it on the ground. Closing the tank fill valve will result in an increase in residual intake pressure because water is no longer flowing through that discharge. Conversely, during periods of low water flow or no water flow, the operator must recognize the need to recirculate water to prevent damage to the pump and/or a loss of prime.

Discharges

Discharges come in a wide variety of sizes and configurations, and the minimum flow ratings of individual discharges are based on the size of the discharge plumbing. According to *NFPA 1900*, pumpers must be equipped with enough 2½" or larger discharges to flow the pump's rated capacity. In addition, the standard requires that any preconnected discharge be plumbed with no less than 2" plumbing.[18]

It is important for the operator to understand that no discharge larger than 2½" is permitted to be at the operator's pump panel.[19] This means that most large diameter discharges will be located either on the passenger's side of the apparatus or at the rear of the apparatus. In an effort to maximize the safety of the pump operator, some departments have designed their pumpers with no discharges on the operator's pump panel (fig. 1–10). While this configuration does create a safer working environment for the pump operator, it also limits the functionality of the pumper when supplying lines for incidents on the driver's side of the rig.

Figure 1–10. This pumper is not equipped with discharges on the driver's side pump panel.

Every discharge on the pumper will have a corresponding discharge valve. Each discharge valve will have a label that corresponds to the individual discharge gauge for that particular discharge. *NFPA 1900* requires that discharges with 3" or larger plumbing be equipped with slow operating valves to minimize the potential for damage to occur from water hammer when these valves are opened or closed.[20] These discharges are typically large diameter discharges, deck guns or "wagon pipes," and aerial discharges. The operating mechanisms for these valves may include hand wheels, cranks, or electric control valves.

The master discharge gauge

The *master discharge gauge* gives the operator crucial information on the discharge side of the pump. This gauge measures pressure as the water leaves the pump, but before it flows through the individual discharge. Because of this, the master discharge gauge will always read the highest pressure leaving the pump. A common mistake that many pump operators make occurs when they are attempting to set the pump pressure for an individual line, and they watch the master discharge gauge rather than the individual line gauge. This causes the line being supplied to be underpumped due to friction loss in the discharge plumbing of the pump (fig. 1–11).

Figure 1–11. The master discharge gauge reads the highest pressure leaving the pump. Note that in this photo, the master discharge gauge is reading 250 psi, while the aerial discharge gauge is reading 200 psi. The difference between the readings represents the friction loss within the plumbing of the apparatus.

The amount of internal friction loss within the pump will vary depending on how the discharge is plumbed, the internal diameter of the plumbing, and the flow rate through the discharge. This is one of the many reasons why it is extremely important for a fire department to flow test their discharges, hose, and nozzle packages on each individual engine. Not only will this help to ensure that operators pump the appropriate pump pressures, but it also will help the department identify which discharges are more advantageous for large flow operations.

Individual discharge gauges

As mentioned above, individual line or discharge gauges measure the pressure of the water for each discharge on the pumper. The gauges read pressure on the outlet side of the discharge valve and thus only read pressure when the valve is opened. *NFPA 1900* does not specify where on the outlet side of the valve the pick-up tube for the individual line gauge must be plumbed. This is quite important to remember because if the gauge pick-up is close to the discharge valve and there is a long run of pipe or numerous bends, the reading will be higher than the pressure of the water at the actual discharge connection. For this reason, it is critically important that fire departments specify on new apparatus that the pick-up tube for individual discharges be plumbed as close to the actual discharge connection as possible. This helps to ensure the most accurate reading.

When charging attack and supply lines, pump operators must remember to set their pump pressures based on the readings of the individual line gauges, *not* the master discharge gauge. This ensures that the appropriate volume of water exits the other end of the attack or supply line. The operator should also remember that the drain valves for each discharge are typically connected to the same tubing as the individual discharge gauge. This allows water to drain out the entire line and prevents the gauge from freezing during cold temperatures.

If a drain valve is accidentally left open, and the corresponding discharge valve is opened to supply a line, the individual discharge gauge will read 0 psi. The result will be an attack or supply line that is charged and flowing water, but a pressure gauge that reads 0 psi. Often the operator just assumes the gauge is damaged. If this situation occurs, the operator should first look under the apparatus to see if water is flowing out of the discharge drain valve. If so, the operator should close the corresponding discharge drain valve on the pump panel. The result should be an immediate reading on the discharge gauge supplying the hose line.

Automatic Pressure Control Devices

The potential for pressure surges on the discharge side of the pump can cause injury to firefighters and damage to equipment and infrastructure. For these reasons, pressure control devices are required on the discharge side of the apparatus. *NFPA 1900* requires that an automatic pressure control device be installed on pumping apparatus that prevents a rise in discharge pressure of more than 30 psi within 3 to 10 seconds of the rise in pressure.[21] On modern apparatus, these devices take the form of either a *discharge relief valve* or a pressure governor.

Discharge relief valves

Discharge relief valves are designed to relieve excessive pressure on the discharge side of the pump (fig. 1–12). They operate in a similar fashion to intake relief valves, as discussed earlier in this chapter. The operator must set the discharge relief valve to the line requiring the highest discharge pressure. This line will have its discharge valve fully opened, while all other lines with lower pressure requirements will have to be gated back. If any of the lines shut down, and the discharge pressure rises above the set level, the discharge relief valve will open and dump the excess pressure either to the intake side of the pump or to the atmosphere, depending on the style of valve installed. In order to comply with the requirements of *NFPA 1900*, all relief valves must have an indicator to signal to the pump operator that the valve is open and is relieving pressure.[22]

Figure 1–12. A common discharge relief valve installed on modern pumpers

Pump operators must remember that if they place a line in service that requires a higher pump pressure, the discharge relief valve must be reset to accommodate the higher pump pressure. All of the other discharges will have to be gated back to their respective pump pressures. If they fail to set the relief valve to a higher pressure and simply attempt to throttle up, the discharge relief valve will open, and pressure will not build past the relief valve setting.

One important thing to remember with regard to discharge relief valves is that they do not manipulate the throttle of the apparatus like a pressure governor does. This means that the throttle always stays at the revolutions per minute set by the operator. This is why setting the discharge relief valve is so critically important. If the discharge relief valve is not set and an attack line shuts down, the engine speed will remain the same, and the result is an increase in pressure on the other lines. A similar scenario would occur if the operator were to open an intake from a pressurized source without first setting the discharge relief valve. This is why when acting as the attack pumper, the pump operator should charge the initial attack line to its appropriate pressure, set the discharge relief valve, and then open the intake.

A common mistake that operators make is *chasing the relief valve.* This typically occurs when a pressure surge causes the relief valve to open, either because a pressurized supply has been received or the operator shuts down a discharge feeding a line. Many operators feel the need to increase the relief valve setting in order for the relief valve to close. While this closes the relief valve, it also increases the pressure setting of the relief valve and will potentially overpressurize the remaining lines in service. The operator should instead lower the engine throttle in these instances. This will help to lower the *net pump discharge pressure* to the setting of the discharge relief valve, thus causing it to close. Remember, the discharge relief valve establishes the highest pressure that can be built on the discharge side of the pump, and the throttle is the means by which pressure is built. Once the discharge relief valve is set, it should not need to be readjusted unless the highest pressure requirement changes.

Pressure governors

Pressure governors have become extremely popular on modern fire apparatus for controlling pump discharge pressures. These devices essentially utilize a computer and pressure transducers to allow the operator to either maintain a set pump discharge pressure or a specific level of revolutions per minute for the engine (fig. 1–13). When operating an apparatus equipped with a pressure governor, the pump operator must recognize which mode the pressure governor should be in based on the role of that particular pumper. The two modes that pressure governors operate in are either the *pressure mode* or the *rpm mode.*

In the pressure mode, the pressure governor manipulates the revolutions per minute of the engine to maintain a specific pump discharge pressure. For example, if the pump operator sets the pressure governor to 150 psi, the governor will automatically adjust the engine revolutions per minute up or down to maintain this pump pressure. When an attack line opens up, the governor will respond to a drop in pressure by increasing the engine revolutions per minute to maintain the pump pressure of 150 psi. If the attack line were to shut down, the governor would respond to the increase in pressure by lowering the engine revolutions per minute to bring the pump pressure back down to 150 psi.

When pump operators are preparing to switch from their tank water to an external pressurized source, the pressure governor in the pressure mode will assist in preventing excessive pressure spikes. After connecting the

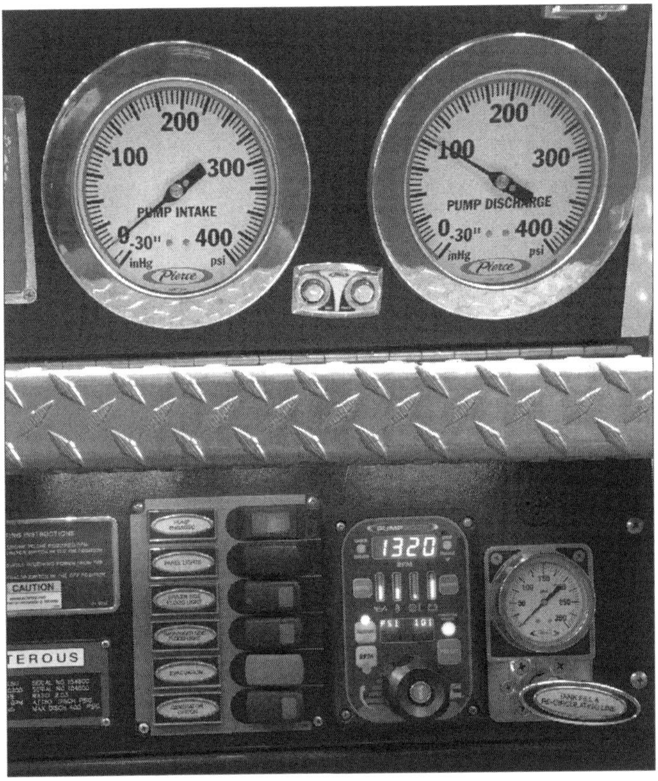

Figure 1–13. A pressure governor operating in the pressure mode. The governor is set to maintain a pump discharge pressure of 101 psi and will do so by increasing or decreasing the engine revolutions per minute as lines open or close.

supply line to a closed intake, charging the supply line, and bleeding the air, the operator should slowly open the intake valve. When the pressurized water enters the pump, the pressure governor will sense an increase in discharge pressure and begin throttling down to maintain the desired pump discharge pressure. For this reason, pump operators must open the intake valve slowly. Failure to open the intake valve slowly will result in the pressure increasing faster than the pressure governor can compensate for.

It is also important to note that when operating from a pressurized water supply, the pressure governor can only lower the engine revolutions per minute of the pumper to idle. If the water source is providing the pumper with a significant amount of incoming pressure, even with the engine idling, there could still be too much pressure on the discharge side of the pump. In this scenario, the pump operator may have to gate back the discharges to ensure the lines being supplied are not overpressurized. In this case it may be advantageous to set the apparatus intake relief valve to a lower setting to help compensate for the high incoming pressure.

To protect the pump, most pressure governors are programmed to respond in a particular fashion to low or absent water supply situations. The three most common occurrences are known as *running away from water*, *low water supply*, and *no water supply*. Running away from water commonly refers to a scenario whereby an increase in engine revolutions per minute does not result in an increase in pump discharge pressure. If the operator continues to increase the throttle, *cavitation* is likely to occur.

In the pressure mode, when the pressure governor registers a drop in discharge pressure, it responds by increasing the engine revolutions per minute. If there is no corresponding increase in the discharge pressure, the pressure governor recognizes this as a running away from water condition. It will stop attempting to throttle up and will signal a warning message to the operator.[23] The solution to this scenario typically would be for the governor to recognize an increase in pressure back to the original discharge setting. Each brand and model of pressure governor will specify remedies to this situation, however.

A low water supply situation occurs when the pressure governor senses a drop in discharge pressure within a specific low-pressure range. One pressure governor on the market enters the low water cycle when the discharge pressure drops below 45 psi but remains above 15 psi. When the governor recognizes this situation, it raises the throttle to a preprogrammed setting for engine revolutions per minute and displays a message to notify the operator of the situation. If the pressure does not rise above 45 psi within a specific period of time, the pressure governor will automatically idle the engine.[24] The specific time period and pressure ranges may vary between apparatus, and some pressure governors can be reprogrammed to meet the needs of a particular department.

A no water supply condition occurs when the discharge pressure drops below the bottom value of the low water cycle. For the example prior, the pressure governor in the pressure mode would enter the no water condition when the discharge pressure dropped below 15 psi.[25] In this scenario, the pressure governor will immediately bring the engine revolutions per minute down to idle. If the water supply is reestablished within a preprogrammed amount of time, the pressure governor will either enter into a low water condition or return to the original pressure setting. This depends entirely on the model of governor and what discharge pressure results from the reestablished water supply. If the water supply is not reestablished within the programmed time period, the governor resets.

If the pump operator is acting as the attack engine, the pressure governor should be set in the pressure mode. This ensures that the discharge pressure remains at the required setting as attack lines open and close. The operator must remember that in the pressure mode, the governor is set to the line requiring the highest discharge pressure; all other lines will be gated back. The governor will prevent pressure surges on all the discharges operating off the pumper.

In the rpm mode, the pressure governor will maintain a specific setting for engine revolutions per minute. Older pressure governors did this regardless of decreases or increases in pressure. If the operator set the governor to 1,500 rpm, the governor would maintain that setting no matter what. On modern apparatus, the pressure governor in the rpm mode will adjust the engine revolutions per minute if it senses a discharge pressure increase of 30 psi over the last established discharge pressure value.[26] This safety feature is designed to meet the requirements in *NFPA 1900* for automatic pressure control devices.[27]

When operating modern pumping apparatus, the operator should remember that the pressure governor in the rpm mode does not respond to a drop in discharge pressure. If the discharge pressure drops all the way to 0 psi, the governor will still maintain the rpm value set by the pump operator. This is why the operator should be in the rpm mode when drafting. If the operator attempts to draft in the pressure mode, the governor will enter the no water cycle and prevent the operator from throttling up to achieve the necessary engine revolutions per minute for the priming operation.

Pressure governors in newer apparatus only respond to an increase in discharge pressure above 30 psi. In this instance, the pressure governor will lower the engine revolutions per minute to bring the discharge pressure back to the last pressure setting. The governor will also notify the operator of this condition either by a message on a display or by a flashing light. Common conditions that could cause the discharge pressure to increase 30 psi or more would include a significant boost in intake pressure or shutting down a heavy-caliber master stream. A 30 psi jump in pressure typically does not occur with the opening and closing of hand lines. It is for this reason that the pump operator should be in the rpm mode whenever operating as a supply pumper or a relay pumper.

The pump operator must remember that in the rpm mode, the governor does not respond to decreases in pressure. This means that if the governor lowered the engine revolutions per minute due to a 30 psi increase in pressure from a ladder pipe shutting down, the governor will *not* increase the engine revolutions per minute once the ladder pipe begins flowing again. It is the responsibility of the pump operator to watch the gauges and realize the necessity of throttling back up to the required pressure when the discharge pressure drops on the line feeding the ladder pipe.

Study Questions

1. Which of the following is a variable that impacts the rated pump capacity of a centrifugal fire pump?

   a. The size of the pump's impeller
   b. The number of pump discharges
   c. The horsepower of the vehicle's motor
   d. All of the above

2. Centrifugal fire pumps are rated to flow 70% of their pump capacity at what pressure?

   a. 150 psi
   b. 200 psi

c. 250 psi
d. 300 psi

3. Centrifugal fire pumps are designed to accomplish which of the following?

a. Flow high volumes of water at high pressures
b. Flow low volumes of water at low pressures
c. Flow high volumes of water at low pressures
d. None of the above

4. For pumpers with 500-gallon booster tanks or larger, what is the minimum tank-to-pump flow rate that the plumbing and valve assembly must be capable of providing according to *NFPA 1900*?

a. 250 gpm
b. 500 gpm
c. 750 gpm
d. 1,000 gpm

5. Which of the following gauges on the pump panel is required to be a compound gauge according to *NFPA 1900*?

a. The master discharge gauge
b. The master intake gauge
c. All pressure gauges
d. The tachometer

6. Pump discharge pressures should be set based on which gauge?

a. The master discharge gauge
b. The master intake gauge
c. The compound gauge
d. The individual line gauge for the particular discharge

7. Automatic pressure control devices must prevent the discharge pressure from rising more than 30 psi within ____________.

a. 1 second
b. 3–10 seconds
c. 10–15 seconds
d. 30 seconds

8. Which of the following modes can be found on a pressure governor?
   a. Parallel
   b. Series
   c. Volume
   d. Rpm
9. The pressure governor should be placed in the pressure mode when the apparatus is serving as the ___________ engine.
   a. Attack
   b. Supply
   c. Relay
   d. Both B and C
10. The pressure governor should be placed in the rpm mode when the apparatus is serving as the __________ engine.
   a. Attack
   b. Supply
   c. Relay
   d. Both B and C

2

HYDRAULICS FOR THE SUPPLY PUMPER

Hydraulic calculations are arguably the most dreaded responsibility of the pump operator. When we look at hydraulic calculations from the standpoint of the attack engine, however, only a minimal amount of math must be done when pumping a preconnected attack line. These pump pressures should already be known by the operator of the attack engine. This is accomplished by performing hose and nozzle tests for each preconnected line on the apparatus and developing a pump chart or committing these values to memory. Therefore, there are only a handful of scenarios in which hydraulic calculations must be performed on scene due to changing variables. These scenarios include the following:

- Extended lay or leader line operations
- Portable master stream operations
- Standpipe operations
- Elevated master stream operations
- Relay pumping or supply operations

Of these scenarios, two stand out as evolutions that will require hydraulic calculations to be performed by the supply pumper—elevated master stream operations and relay pumping operations. Pump operators must be proficient with pumping water for all of the evolutions listed above, but we will focus specifically on the hydraulics associated with supply engine operations. For this reason, the reader should have a baseline understanding of the principles relating to hydraulic calculations. In this chapter, we will introduce concepts that will make determining the appropriate pump discharge pressures for water supply operations simple, rapid, and accurate.

The Master Equation

When the topic of hydraulics is discussed, there is usually a whirlwind of various equations that enter the pump operator's mind. While all of the equations taught in a pump operator's initial operator training program are important, the *pump discharge pressure equation* is perhaps the most important. This "master equation" is utilized by the pump operator to determine the required pump discharge pressure for the particular evolution being supplied. The equation to determine pump discharge pressure (*PDP*) is expressed as follows, with all units expressed in pounds per square inch of pressure:

$$PDP = NP + TFL + AL \pm EL$$

where

NP = Nozzle pressure
TFL = Total friction loss
AL = Appliance loss
EL = Elevation loss or gain

Regardless of the pump scenario, the operator should always account for all four of these variables. This will ensure that the pump operator is overcoming all of the potential losses in the system and is supplying the appropriate volume of water to the line connected to the rig. If one of the variables is not applicable, such as when there is no appliance in the hose layout, the operator should simply assign "0 psi" for that variable. During water supply operations, the PDP equation will be the basis for determining the correct pump discharge pressure for a particular supply evolution. In the sections that follow, we will break down each of the variables of this equation as they relate to the supply pumper.

Nozzle pressure and residual intake pressure

Nozzle pressure can be defined as the pressure required at the nozzle to produce an effective fire stream and flow its rated capacity. This variable is often considered during attack engine operations, and the value applied will depend on the type of nozzle being utilized. Firefighting nozzles are broken into two broad categories: *smoothbore nozzles* and *combination fog nozzles*. Nozzles are further classified as either *handline* or *master stream nozzles* depending on the volume of water they are designed to flow. Handline nozzles are considered to flow less than 350 gpm, while master stream nozzles are capable of flowing more than 350 gpm. General nozzle pressure values used in the fire service are as follows:

- 50 psi for smoothbore handlines
- 80 psi for smoothbore master streams
- 100 psi for combination fog nozzles, handlines, and master streams

The reader must understand these are general values used in the fire service. Some fire departments elect to overpump or underpump their smoothbore handline nozzles based on the flows desired and the associated nozzle reaction produced. During high-velocity master stream operations, smoothbore master stream nozzles are pumped at nozzle pressures well above 80 psi. These operations and their effects on aerial apparatus will be discussed in chapter 8. It is also important to recognize that some nozzle manufacturers produce combination fog nozzles with nozzle pressures of 75 psi or even 50 psi. Pump operators must be knowledgeable concerning the nozzle types and the associated nozzle pressures for each piece of equipment used by their organization.

Pumping to the intake of another engine requires the operator to take a slightly different approach with regard to nozzle pressure. In these situations, there is technically no nozzle attached to the supply line that the supply engine is feeding. This causes many operators to assign a value of 0 psi for the nozzle pressure variable in the PDP equation. Doing so will cause their pump discharge pressure calculation to be too low, however.

Whenever a supply pumper is feeding the pump of another apparatus, the operator must be mindful of providing an adequate *residual intake pressure (IP)* for the engine being supplied. Residual intake pressure is referred to as the pressure remaining in a system once water is flowing. The residual intake reading is the remaining pressure on the master intake gauge as a pumper is discharging water. When determining the required pump discharge pressure for a relay operation, the pump operator must account for the residual intake pressure on the next pumper in addition to all the other loss values for the hose layout. This will allow a positive intake reading on the next engine's master intake gauge. Failure to account for the necessary residual intake pressure on the next pumper could result in cavitation or a collapse of the supply line.

An acceptable range for the residual intake pressure when calculating the pump discharge pressure for a supply operation is considered to be between 20 psi and 50 psi. This allows the next pumper in the operation to have a sufficient amount of intake pressure to keep their supply line from collapsing while also preventing overpressurization on the next pumper's intake. The goal of the supply pumper and each of the relay pumpers (if present) is to maintain this residual intake pressure range for each subsequent engine involved with the operation.

Pump operators can think of the value assigned for residual intake pressure as being similar to the nozzle pressure (*NP*) for the operation. Therefore, it is acceptable to substitute *IP* for *NP* in the pump discharge pressure equation. Doing so allows us to rewrite the equation to determine pump discharge pressure (*PDP*) with one slight modification, with all units expressed in pounds per square inch of pressure:

$$PDP = IP + TFL + AL \pm EL$$

where

IP = Residual intake pressure
TFL = Total friction loss
AL = Appliance loss
EL = Elevation loss or gain

Essentially, the next pumper in the relay serves as the "nozzle" at the end of the supply line. The difference here is that there is a wider range of acceptable "nozzle pressures." Regardless of how this is visualized, the operator must account for residual intake pressure when determining the required pump pressure for a supply operation. This value will then be added to the friction loss in the supply line along with any appliance loss or elevation loss or gain present to determine the required pump discharge pressure.

Total friction loss

Whenever water moves through hoses, pipes, appliances, and fittings, there will be friction that will slow the water down. This loss of energy is known in the fire service as *friction loss*, and it is expressed in pounds per square inch (psi). *Total friction loss* is defined as the sum of all the friction loss values within a hose layout. For a single preconnected handline scenario, this is simply the friction loss (*FL*) in the hose that is attached to the nozzle.

In a scenario where there is a 2½" line feeding a 1¾" line via a reducer, the total friction loss (*TFL*) value would be the friction loss in the 2½" line plus the friction loss in the 1¾" line.

The pump operator must have a solid understanding of the following rules governing friction loss:[1]

1. *Friction loss varies with the quality of the hose.* As fire hose ages, its lining deteriorates. This has a direct effect on the friction loss characteristics of a particular hose line. It is also one of the reasons why it is recommended that fire departments establish a hose replacement schedule. The operator must also understand that different hose manufacturers utilize different manufacturing processes to produce their fire hoses. This includes various materials and even different weave patterns on the hoses. All of these factors will influence the amount of friction loss for a particular brand of hose. For these reasons, the pump operator should remember that hoses will vary.
2. *Friction loss varies directly with the length of the hose.* In the U.S. fire service, friction loss values are typically expressed per 100' of hose. This rule states that if the hose length is doubled, the friction loss in the system will also double. The operator should remember that friction loss is directly proportional to an increase in hose length.
3. *Friction loss varies with the square of the velocity.* This principle has the biggest impact on friction loss. In order to break this statement down into manageable parts, picture water flowing through a 1¾" hose line. If we were to increase the velocity of the water moving through that 1¾" line, the result would be an increase in the flow rate out of the nozzle. This means that when the area remains constant, an increase in velocity results in an increase in flow rate. What this rule essentially tells us is that friction loss in a system increases much faster than an increase in the flow rate. To be more specific, when the flow rate through a hose line doubles, the friction loss will quadruple. The pump operator must recognize that the flow rate has the biggest effect on the friction loss within that system. This is because friction loss is exponentially proportional to an increase in the flow rate.
4. *For a given flow rate, friction loss varies inversely as the fifth power of the hose diameter.* The previous rule stated that when the hose diameter or area remains the same, friction loss increases much faster than an increase in the flow rate. The fourth rule of friction loss

explains that the most efficient way to reduce friction loss is to increase the hose diameter. Simply put, if you want more water with less friction loss, use a larger diameter hose. Hence the advent of the use of *large diameter hose (LDH)* in the modern fire service. It is important to remember that it may not always be possible to simply switch to a larger diameter hose, however. During some supply operations, it may be necessary to flow more water than a single 4" or 5" large diameter supply line can support. For example, suppose we are supplying water to a fire scene utilizing a single 4" supply line. If we wanted to double the flow rate at the fire scene from 1,000 gpm to 2,000 gpm, we know that the friction loss in the single 4" supply line would quadruple. We also know that we could keep the friction loss lower while flowing 2,000 gpm if we switch out the supply line with 6" hose. This is not a feasible option for the majority of the municipal fire service, however. Instead, the supply engine can perform what is known as a *parallel supply line operation* by laying a second 4" supply line to the fire scene (fig. 2–1). When we perform this operation, the total 2,000 gpm flow rate is split between each 4" supply line, and each line is moving 1,000 gpm. By doing this, the friction loss in each of the 4" supply lines moving 1,000 gpm is the same as what the friction loss would be in a single 6" supply line moving 2,000 gpm. Although we did not physically put a 6" supply line on the ground, by performing a parallel 4" supply line operation, we reaped the same benefits hydraulically as

Figure 2–1. Feeding this elevating platform with two 4"supply lines is the hydraulic equivalent of feeding it with a single 6" supply line.

if we did lay a 6" supply line. Parallel supply line operations will be discussed in more detail in later chapters.

5. *For a given velocity, friction loss is independent of the pressure.* It is important to remember that friction loss is primarily dictated by the flow through a given length of hose. It is not based on the pressure of the water. If we are on level ground and flowing 1,000 gpm through 300' of 4" supply line, the friction loss would be 60 psi. If that same supply line were going 100' up a hill, the friction loss in the supply line would remain the same, but the pump discharge pressure at the pump panel would have to be higher to compensate for the elevation loss going up the hill. In both cases, the friction loss remains the same, even though the pump pressures at the panel would differ. Remember, friction loss is based on hose diameter, hose length, and flow rate.

There are many ways to determine the friction loss in a hose layout. One of the most accurate methods is the *coefficient formula*. This method applies each of the rules listed above in the form of a mathematical expression. The coefficient formula for determining friction loss (FL) is expressed as follows:

$$FL = CQ^2L$$

where

C = Coefficient of friction
Q = Flow rate in gallons per minute ÷ 100
L = Hose length in feet ÷ 100

The hose coefficient assigned to a particular hose line provides the pump operator with crucial information about the diameter of that hose and how smooth its lining is. The higher the coefficient value, the smaller the hose diameter and/or the rougher the inner lining of the hose. Several sizes of hose are utilized by fire departments across the United States for supply operations. Table 2–1 lists the coefficients for hoses commonly used as supply lines.[2]

The pump operator must understand that different hose manufacturers assign different hose coefficients to their products. This is partially because of variations in manufacturing techniques but can also be due to the fact that some manufacturers produce hoses that have larger internal diameters than advertised. While there have been advances in supply hose over the

Table 2–1. Coefficients of common supply lines

Supply hose diameter	Coefficient
2½"	2
3" with 2½" couplings	0.80
4"	0.20
5"	0.08
6"	0.05

Source: C. Brakhage and L. A. Miller, *Pumping and Aerial Apparatus Driver/Operator Handbook*, 3rd ed. (Stillwater, OK: Fire Protection Publications, OSU, 2015), 222.

years that have resulted in lower coefficient values, it is generally acceptable to use the values listed in table 2–1 for calculating the friction loss in supply lines. The calculations performed in this text are based on these hose coefficients.

When discussing fire service hydraulics, Q always represents the flow rate within a system. As discussed previously, the flow rate in gallons per minute has the greatest effect on the friction loss in a hose layout. This is evident when looking at the coefficient formula because Q is the only variable in the equation that is squared, proving that when the flow rate is doubled, the friction loss within the system will quadruple. This is one of the many reasons why knowing the flow rate in a hose layout is so important for the pump operator. The coefficient formula also proves that there is a linear relationship between friction loss and hose length. As the hose length doubles, so will the friction loss. Examples 2–1 and 2–2 highlight the effect on friction loss of doubling the flow rate when the hose diameter and length remain constant.

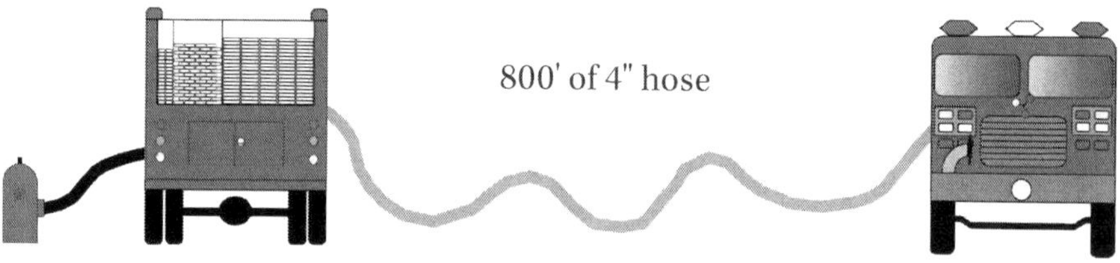

Example 2–1. What is the friction loss in 800' of 4" hose flowing 500 gpm?

$FL = CQ^2L$

$FL = 0.2\,(5)^2\,8$

$FL = 0.2\,(25)\,8$

$\mathbf{FL = 40\ psi}$

Example 2–2. What is the friction loss in 800' of 4" hose flowing 1,000 gpm?

$FL = CQ^2L$
$FL = 0.2\ (10)^2\ 8$
$FL = 0.2\ (100)\ 8$
$\boldsymbol{FL = 160}$ **psi**

Example 2–3 highlights the effect on friction loss within the system when larger supply hose is used for the operation. In this example, the flow rate and hose length are the same as in example 2–2, but the supply line utilized is 5" LDH.

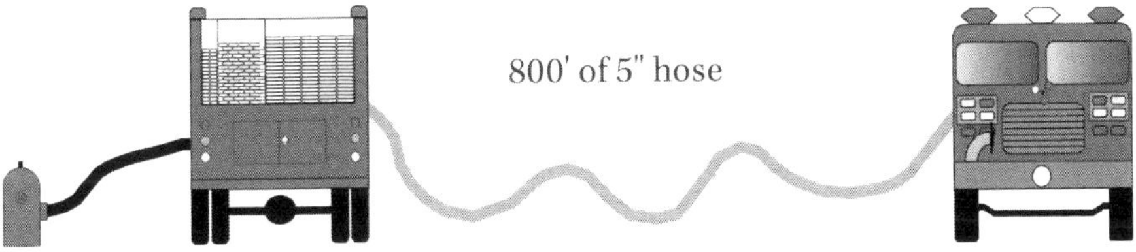

Example 2–3. What is the friction loss in 800' of 5" hose flowing 1,000 gpm?

$FL = CQ^2L$
$FL = 0.08\ (10)^2\ 8$
$FL = 0.08\ (100)\ 8$
$\boldsymbol{FL = 64}$ **psi**

This example proves that increasing the hose diameter has a significant impact on the friction loss within a supply line layout. By switching from 4" hose to 5" hose in this example, the friction loss in the system is reduced by 60%. It is not always feasible to switch from 4" hose to 5" hose in the middle of an operation, however. In example 2–2, a single 4" supply line is laid 800' to the scene, and the supply pumper is attempting to deliver 1,000 gpm to the attack engine. Example 2–4 evaluates the effect on the friction loss within the system resulting from laying an additional 800' of 4" supply line between the source and the scene.

In order to solve this problem, we first have to determine what the total flow rate at the fire scene is. In this example, the fire scene is flowing 1,000 gpm. Since we have two supply lines feeding the scene, that means that 500 gpm will be flowing in each supply line. Both of these supply lines are the same length and diameter and are flowing the same volume of water, so we only have to do the math for one of the lines. The friction loss will be the same value in the other line. In examples 2–2 and 2–4, the fire scene is being supplied with 1,000 gpm flow. The use of parallel 4" supply lines in example 2–4, however, resulted in a 75% reduction in friction loss in the system.

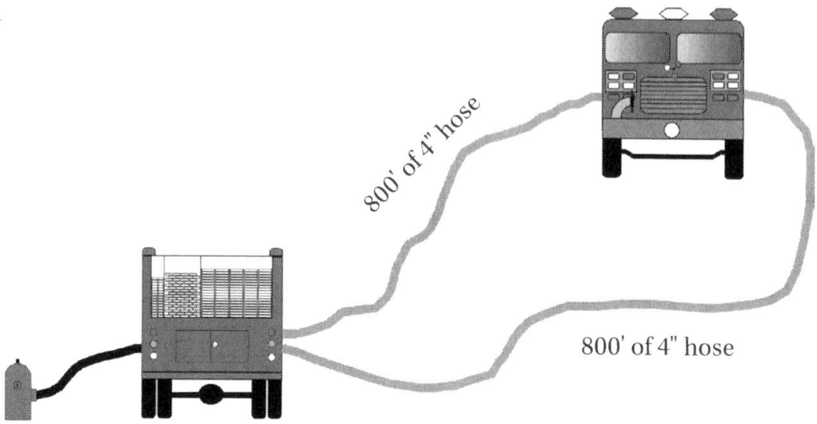

Example 2–4. You are performing a parallel supply line operation utilizing two 4" supply lines that are 800' in length. The fire scene requires 1,000 gpm. What is the friction loss in each of the supply lines?

$$FL = CQ^2L$$
$$FL = 0.2(5)^2 8$$
$$FL = 0.2(25)8$$
$$\mathbf{FL = 40\ psi}$$

Appliance loss

Up to this point, we have discussed friction loss as it pertains to fire hose. The operator must remember that there is also friction loss within the appliances used in common hose layouts, which is referred to as *appliance loss*. While there is no particular calculation for determining appliance loss, the rules governing friction loss also apply to appliances used in a hose layout. For example, an appliance with a larger waterway will have less appliance loss at a given flow rate than an appliance with a smaller waterway.

Most driver operator training programs teach firefighters to account for appliance loss based on the appliance being utilized and the flow rate through the appliance. Appliances are generally broken into three categories: *general hose appliances*, *master stream devices*, and *fire department connections (FDCs)*. General hose appliances include gated wyes, siamese appliances, water thieves, and manifolds, to name a few. In most texts, the appliance loss through general appliances is dependent on the flow rate through the appliance. If the flow rate through the appliance is less than 350 gpm, the appliance loss is considered to be 0 psi. In scenarios where the flow rate through the appliance is 350 gpm or greater, the operator should add 10 psi for the appliance loss. When accounting for appliance loss in FDCs and master stream devices, many operators are taught to add 25 psi for these appliances regardless of the flow rate.[3]

These values should be considered general guidelines for determining appliance loss. Operators should remember that appliance loss values may differ among appliance brands and at different flow rates. Some manufacturers will list the appliance loss values for their equipment at different flow rates in the operator's manual for that particular appliance. If this information is not readily available, flow tests should be conducted to determine the actual appliance loss values for the equipment used by the fire department. Chapter 8 will discuss in more detail how to account for appliance loss when flowing more than 1,000 gpm through aerial devices.

Elevation loss or gain

The final variable the pump operator must take into account when determining the pump discharge pressure is the presence of *elevation loss or gain*. In the fire service, we describe *elevation* as the difference in height between the pump and where water is being discharged. The discharge opening is typically a nozzle, but it could be another pumper as well. In either case, gravity will act on the water in the hose line and thus slow it down. The pump operator will have to pump at a higher discharge pressure in order to overcome the forces of gravity. This is known as elevation loss.

Elevation gain occurs when the nozzle or discharge point is located at a point below the pumper. In these scenarios, if the pump operator does not account for the forces of gravity, the line at the bottom of the hill will have too much pressure. When a hose line is located at an elevation below the pumper, the operator must determine what the elevation pressure gain is and subtract it from the other variables in the pump discharge pressure equation. When pumping up or down a hill or pumping up an aerial device, the following formula can be used to determine the elevation loss or gain (EL):

$$EL = H/2$$

where

H = height in feet

The need to supply aerial devices offers a common situation in which the pump operator must determine elevation loss in the field. Operators must remember that when supplying aerial devices, the elevation is measured from where the aerial tip is in relation to the pumper, not the extension of the aerial device (figs. 2–2 and 2–3). Remember, the aerial device could be extended 100', but the elevation of the tip where water is being discharged may be much lower.

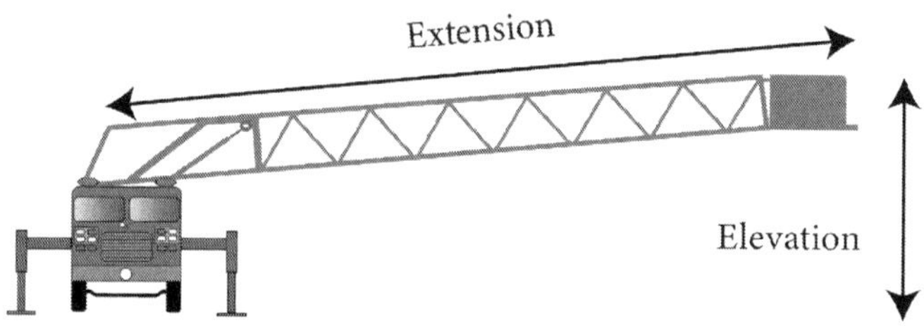

Figure 2–2. An aerial device extension does not directly correlate to elevation height. The aerial device in this example is fully extended, but it is not raised at a high angle. This means the elevation loss is minimal.

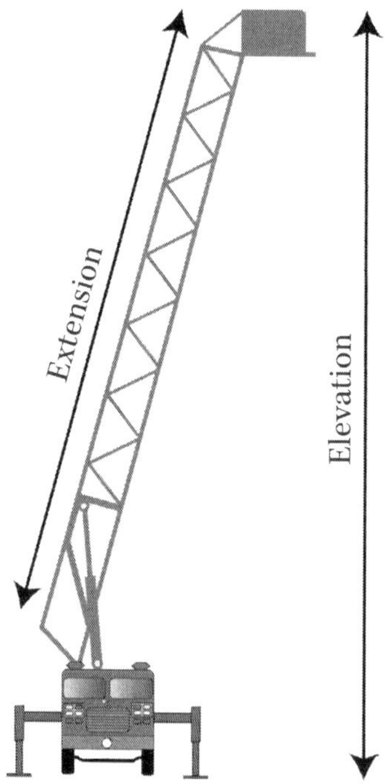

Figure 2–3. As the angle of the aerial device is increased, extension and elevation have a greater correlation.

Application for supply operations

Now that we have discussed the four variables that make up the pump discharge pressure equation, we can utilize them to determine the pump discharge pressures for various supply scenarios. The operator should remember that each of these examples will yield a theoretical answer.

Example 2–5. You are serving as the supply pumper in a relay operation. There is 800' of 4" supply line between your supply pumper and the next relay engine. You are supplying 500 gpm to the next relay engine. What is the required pump discharge pressure of the supply pumper for this operation?

$$PDP = IP + TFL + AL \pm EL$$
$$PDP = (20 \text{ to } 50) + TFL + 0 \pm 0$$
$$PDP = (20 \text{ to } 50) + 40 + 0 + 0$$
$$\mathbf{PDP = 60 \text{ to } 90 \text{ psi}}$$

$$FL_{4"} = C(Q^2)L$$
$$FL_{4"} = 0.2\,(5^2)\,8$$
$$FL_{4"} = 0.2\,(25)\,8$$
$$\mathbf{FL_{4"} = 40 \text{ psi}}$$

Example 2-5 shows how to appropriately calculate the pump discharge pressure requirement for a supply operation. In this example, if the supply pumper sets the pump discharge pressure to 60 psi, the resulting residual intake pressure on the relay pumper would be 20 psi. If the supply engine pumps at a PDP of 90 psi, however, the relay engine will have a residual intake pressure reading of 50 psi, assuming there is no change in elevation. As long as the pump discharge pressure of the supply engine is within this range, the relay engine will be receiving the necessary volume of water with an appropriate amount of residual intake pressure.

Since residual intake pressure has a 30 psi range of acceptable values, the operator has to determine when the high and low ends of the range are warranted. In example 2–5, because the friction loss value is relatively low, it is recommended to apply 50 psi for the residual intake pressure, thus making the pump discharge pressure of the supply engine 90 psi. One advantage to applying a higher residual intake value is that it will be easier to maintain a pump discharge pressure of 90 psi on the supply pumper compared to 60 psi. This is especially true if the supply engine is operating from a hydrant.

Pumping at the higher end of the residual intake pressure range will obviously result in a higher residual intake pressure on the next engine. Not only will this reduce the workload of the relay pumper, but it will also allow the next engineer to operate with greater redundancy on the intake side of the pump. Should the flow rate increase without proper communication, the higher residual intake pressure will help prevent a complete collapse of the supply line. Pumping at the higher end of the residual intake pressure range also aids supply operators in overcoming any other loss values that they either forgot or did not fully account for. Even so, there are situations where the pump operator should pump at the lower end of the residual intake pressure range.

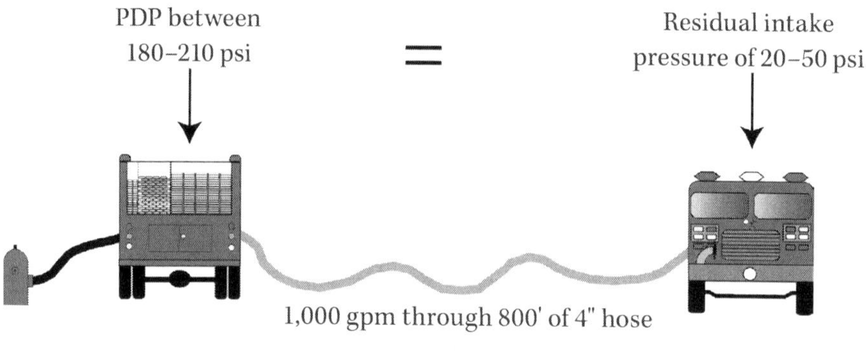

Example 2–6. You are serving as the supply pumper in a relay operation. There is 800' of 4" supply line between your supply pumper and the next relay engine. You are supplying 1,000 gpm to the next relay engine. What is the required pump discharge pressure of the supply pumper for this operation?

$$PDP = IP + TFL + AL \pm EL$$
$$PDP = (20 \text{ to } 50) + TFL + 0 + 0$$
$$PDP = (20 \text{ to } 50) + 160 + 0 + 0$$
$$\mathbf{PDP = 180 \text{ to } 210 \text{ psi}}$$

$$FL_{4''} = C(Q^2)L$$
$$FL_{4''} = 0.2\,(10^2)\,8$$
$$FL_{4''} = 0.2\,(100)\,8$$
$$\mathbf{FL_{4''} = 160 \text{ psi}}$$

The flow rate in example 2–6 doubled compared to example 2–5, while all other variables remained the same. By doubling the flow rate, the friction loss in the supply hose quadrupled. After adding the residual intake pressure range to the friction loss value, the necessary pump discharge pressure for this operation is between 180 psi and 210 psi. In this scenario, it is recommended to apply the lower residual intake pressure value to ensure that the pump discharge pressure does not exceed the pressure rating of the supply hose. If the supply pumper were drafting, less of the rated capacity of the pump would be sacrificed by applying around 20 psi for the residual intake pressure compared to applying 50 psi. When accounting for residual intake pressure on the fireground, it is recommended to calculate the friction loss in the supply line first. This will allow the operator to apply the appropriate residual intake pressure based on all other loss factors.

Fireground Friction Loss Methods for Supply Operations

Calculating friction loss and determining pump discharge pressures on the fireground should be a rapid task for the seasoned pump operator. On the fireground, the pump operator still needs to account for all four variables of the PDP equation. Nozzle pressure/residual intake pressure, appliance loss, and elevation loss or gain are three of the four variables that can be determined fairly quickly. Determining the friction loss of a hose layout on scene can be a bit more difficult. In this section, we will discuss popular fireground methods for determining friction loss in various supply lines.

Condensed Q method

The *condensed Q* method is extremely popular for determining the friction loss in various supply lines. This method can be used to determine the friction loss per 100' in 2½", 3", 4", and 5" hose. As with the coefficient method, *Q* in each of the following equations is represented as the flow rate divided by 100.

While not the ideal supply line of choice, 2½" hose is still used by some fire departments as their primary supply line. In addition, 2½" hose may be utilized during more advanced pump operations for building parallel supply lines and performing *high-volume series pumping operations.* These operations will be discussed in more detail in chapter 9. The following equation is used to determine the friction loss per 100' in 2½" hose ($FL_{2.5'}/100'$) using the condensed Q method:

$$FL_{2.5'}/100' = 2Q^2$$

where

Q = flow rate in gallons per minute

Example 2–7. Utilizing the condensed Q method, determine the friction loss in 100' of 2½" hose flowing 300 gpm.

$FL_{2.5'}/100' = 2Q^2$
$FL_{2.5'}/100' = 2\,(3)^2$
$FL_{2.5'}/100' = 2\,(9)$
$\boldsymbol{FL_{2.5'}/100' = 18}$ **psi**

If we examine this formula closely, we see that we are essentially performing a variation of the coefficient method. This is due to the fact that the coefficient for 2½" hose is 2. Since we are multiplying 2 by Q^2, and then multiplying that result by the number of 100' hose sections, the answer will be identical to that of the coefficient method.

The condensed Q formula for 3" hose is written as follows:

$$FL_{3'}/100' = Q^2$$

Example 2–8. Using the condensed Q method, determine the friction loss in 500' of 3" hose flowing 300 gpm.

$FL_{3'}/100' = Q^2$
$FL_{3'}/100' = 3^2$
$FL_{3'}/100' = 9$ psi
$TFL_{3''} = 9 \times 5$
$\boldsymbol{TFL_{3''} = 45}$ **psi**

As stated above, the condensed Q method is just another way of thinking about the coefficient method. The coefficient for 3" hose with 2½" couplings is 0.8. In the condensed Q formula, we have simply rounded that coefficient value to 1. That is why the formula is just Q^2 when determining the friction loss per 100' of 3" hose. It is important to note that because the coefficient is rounded up, this method does add 20% more friction loss to the final answer compared to the coefficient method. This is of little concern at flows below 500 gpm. When flowing 500 gpm or more through 3" hose, however, it is recommended that the operator account for this 20% extra loss.

To calculate the friction loss per 100' of 4" supply line ($FL_{4"}/100'$), the following condensed Q formula can be utilized:

$$FL_{4"}/100' = Q^2/5$$

Example 2–9. Using the condensed Q method, determine the friction loss in 600' of 4" hose flowing 1,000 gpm.

$FL_{4"}/100' = Q^2/5$
$FL_{4"}/100' = 10^2/5$
$FL_{4"}/100' = 100/5$
$FL_{4"}/100' = 20$ psi
$TFL_{4"} = 20 \times 6$
$\mathbf{TFL_{4"} = 120}$ **psi**

As with the other two condensed Q methods, the formula for 4" hose is also an adaptation of the coefficient formula. The coefficient utilized for 4" hose is 0.2, which is also the decimal form of the fraction $\frac{1}{5}$. Essentially, this formula has the operator divide the Q^2 value by 5 instead of multiplying it by 0.2. Regardless, the results are friction loss values per 100' that are identical to the coefficient method, making this the preferred option in the field for calculating friction loss in 4" hose.

Condensed Q can also be utilized to determine the friction loss per 100' in 5" supply line. While this method does not produce results identical to the coefficient method, the values are considered close enough to be used on the fireground. The following formula is used to calculate the friction loss in 5" hose using the condensed Q method:

$$FL_{5"}/100' = Q^2/15$$

Example 2–10. Using the condensed Q method, determine the friction loss in 800' of 5" hose flowing 1,000 gpm.

$FL_{5"}/100' = Q^2/15$
$FL_{5"}/100' = 10^2/15$
$FL_{5"}/100' = 100/15$

$FL_{5''}/100' = 6.67$ psi
$TFL_{5''} = 6.67 \times 8$
$\boldsymbol{TFL_{5''} = 53.36}$ **psi**

One of the disadvantages of using this method is that the answers for most common flow rates typically contain decimals. On the fireground, we would just round to the nearest whole number or multiple of 5. Since most of us find it somewhat difficult to divide by 15 without the use of a calculator, the double drop zero method is recommended when calculating friction loss in 5" hose on the fireground.

The double drop zero method for 5" hose

On the fireground, calculating friction loss should be quick and simple. This statement holds true for the majority of the fireground friction loss methods previously discussed, with the exception of the condensed Q equation for 5" hose. The double drop zero method offers the pump operator a quick and uncomplicated way to determine the friction loss per 100' of 5" hose on scene. This method is suitable for flow rates between 500 gpm and 1,500 gpm. The operator should understand that although this method may result in values that vary slightly from the coefficient method, the results are acceptable to use on the fireground.

Utilizing the double drop zero method, the operator must first determine the flow rate through the supply line. Once the flow rate is determined, the operator drops the last two digits of the flow rate (typically zeros). The result is the friction loss per 100' of 5" hose.

Example 2–11. Using the double drop zero method, determine the friction loss in 100' of 5" hose flowing 500 gpm.

$FL_{5''}/100' = 5\cancel{00}$
$\boldsymbol{FL_{5''}/100' = 5}$ **psi**

Using the coefficient method, the answer for example 2–11 would be 2 psi per 100'. The condensed Q method for 5" hose would result in a value of 1.67 psi per 100'. Although the double drop zero method results in a slightly higher friction loss value at 500 gpm, the operator can prevent overpressurization during relay pumping operations by applying a lower residual intake pressure value.

Example 2–12. A supply pumper is feeding the attack engine through 1,000' of 5" hose. The supply pumper is flowing 500 gpm through the supply line. Determine the necessary pump discharge pressure for the supply pumper.

Coefficient method

$PDP = IP + TFL + AL \pm EL$

$PDP = (20 \text{ to } 50) + TFL + 0 + 0$

$PDP = (20 \text{ to } 50) + 20 + 0 + 0$

$\mathbf{PDP = 40 \text{ to } 70 \text{ psi}}$

$FL_{5"} = C(Q^2)L$

$FL_{5"} = 0.08\,(5^2)\,10$

$FL_{5"} = 0.08\,(25)\,10$

$\mathbf{FL_{5"} = 20 \text{ psi}}$

Condensed Q method

$PDP = IP + TFL + AL \pm EL$

$PDP = (20 \text{ to } 50) + TFL + 0 + 0$

$PDP = (20 \text{ to } 50) + 17 + 0 + 0$

$\mathbf{PDP = 37 \text{ to } 67 \text{ psi}}$

$FL_{5"} = Q^2/15$

$FL_{5"} = 5^2/15$

$FL_{5"} = 25/15$

$FL_{5"} = 1.67 \text{ psi}/100'$

$TFL_{5"} = 1.67 \times 10$

$\mathbf{TFL_{5"} = 16.7 \text{ psi}}$

Double drop zero method

$PDP = IP + TFL + AL \pm EL$

$PDP = 20 + TFL + 0 + 0$

$PDP = 20 + 50 + 0 + 0$

$\mathbf{PDP = 70 \text{ psi}}$

$FL_{5"} = 5\cancel{00}/100'$

$FL_{5"} = 5 \text{ psi}/100'$

$TFL_{5"} = 5 \times 10$

$\mathbf{TFL_{5"} = 50 \text{ psi}}$

In example 2–12, we see that the double drop zero method results in a much higher friction loss value than those determined with the coefficient and condensed Q methods. By only adding 20 psi for the residual intake pressure to the friction loss value of the double drop zero method, however, the final pump discharge pressure is within the same range as the coefficient method. Since the double drop zero method results in a higher friction loss value, in reality, the residual intake pressure on the next pumper will actually be closer to 50 psi. Regardless, this method not only simplifies the calculations that must be performed on scene, but it also provides the appropriate flow rate and intake pressure to the next engine.

Example 2–13. A supply pumper is feeding a relay engine through 800' of 5" hose. The supply pumper is flowing 1,000 gpm through the supply line. Determine the necessary pump discharge pressure for the supply pumper.

Coefficient method

$PDP = IP + TFL + AL \pm EL$
$PDP = (20 \text{ to } 50) + TFL + 0 + 0$
$PDP = (20 \text{ to } 50) + 64 + 0 + 0$
$\boldsymbol{PDP} = \mathbf{84 \text{ to } 114 \text{ psi}}$

$FL_{5''} = C(Q^2)L$
$FL_{5''} = 0.08\,(10^2)\,8$
$FL_{5''} = 0.08\,(100)\,8$
$\boldsymbol{FL_{5''}} = \mathbf{64 \text{ psi}}$

Condensed Q method

$PDP = IP + TFL + AL \pm EL$
$PDP = (20 \text{ to } 50) + TFL + 0 + 0$
$PDP = (20 \text{ to } 50) + 53 + 0 + 0$
$\boldsymbol{PDP} = \mathbf{73 \text{ to } 103 \text{ psi}}$

$FL_{5''} = Q^2/15$
$FL_{5''} = (10^2)/15$
$FL_{5''} = (100)/15$
$FL_{5''} = 6.67 \text{ psi}/100'$
$TFL_{5''} = 6.67 \times 8$
$\boldsymbol{TFL_{5''}} = \mathbf{53 \text{ psi}}$

Double drop zero method

$PDP = IP + TFL + AL \pm EL$
$PDP = 20 + TFL + 0 + 0$
$PDP = 20 + 80 + 0 + 0$
$\boldsymbol{PDP} = \mathbf{100 \text{ psi}}$

$FL_{5''} = 1{,}0\cancel{00}/100'$
$FL_{5''} = 10 \text{ psi}/100'$
$TFL_{5''} = 10 \times 8$
$\boldsymbol{TFL_{5''}} = \mathbf{80 \text{ psi}}$

Example 2–13 again proves how the double drop zero method simplifies the hydraulic calculations that must be performed on the fireground while also putting the pump discharge pressure within the acceptable range. Pump operators should remember that when supplying flows between 500 gpm and 1,000 gpm, they should only add 20 psi for the residual intake pressure using this method. Applying a higher intake pressure value could result in a considerably high intake pressure on the next pumping apparatus.

The pump operator must realize that pump discharge pressure calculations associated with supply operations are intended to get the pump discharge pressure within an acceptable range. The calculations are not intended to pinpoint a specific pressure value as with attack lines. There may be instances where the supply operator correctly calculates the pump discharge pressure for the operation, but the next pumper radios to them that the pressure needs to be adjusted. One reason for this could simply be the comfort of the next operator that you are supplying. For example, some operators are hesitant to operate with a residual intake pressure of 20 psi. These operators may radio to the engine supplying them to ask for more intake pressure. As long as increasing the discharge pressure on the supply pumper does not result in overpressurization of the supply line, this request should be granted. If it is not feasible to do so, then inform the receiving pumper of the inability to boost the pressure.

Pump operators should not view a request for more or less pressure in a relay operation as a failure on their part. Often this request is made because there is a variable that the supply pump operator is unable to account for. This could be that the fire scene increased the fire flow without informing all of the apparatus involved in the supply operation. Another culprit could be a change in elevation that the supply or relay pump operators cannot see or are unaware of. These two variables could drastically change the pump discharge pressure requirements of the supply operation.

Determining Discharge Appliance Loss and Maximum Flow Rates

Knowledge of the capabilities of individual discharges will allow for efficient water supply operations, especially during large flow incidents. This requires the department to flow test each of the discharges on their apparatus used for supply operations to determine the internal plumbing loss at various flow rates. Information obtained from these tests can then be utilized to determine the realistic maximum flow rate that each discharge is capable of flowing. This is not the same procedure for flow testing and determining pump discharge pressures for preconnected attack lines.

The following pieces of equipment are required in order to determine the maximum realistic flow rates of discharges used for supply operations:

- A flow test evaluation form
- A flowmeter if combination fog nozzles are utilized
- A pitot gauge if smoothbore nozzles are utilized or to ensure the flowmeter being utilized is calibrated
- Medium and large diameter supply lines to feed various monitors
- Monitors capable of flowing at least 1,000 gpm
- A high-volume pressurized water source

The first step in this process is to identify the plumbing size of each discharge on the apparatus. Using the values listed in *NFPA 1900*, the minimum expected discharge rates for each discharge can be determined based on their diameter. According to the standard, 2½" discharges must flow a minimum of 250 gpm, 3" discharges must flow a minimum of 375 gpm, and 4" discharges must flow a minimum of 625 gpm.[4] Pumpers today are typically equipped with multiple 2½" discharges and at least one large diameter discharge. Operators may not give much thought to the capacity of a large diameter discharge. The typical belief is that large diameter discharges can sufficiently supply anywhere from 1,000 gpm to 2,000 gpm through a single LDH supply line, since mathematically, 4" hose can easily support a 1,000 gpm flow rate, and 5" hose can support a 2,000 gpm flow rate.

It is important to recognize that the NFPA standard does not specify that large diameter discharges must be a certain diameter. Depending on the apparatus specifications and the apparatus manufacturer, large diameter discharges on municipal pumpers are typically plumbed with either 3" or 4" plumbing. According to the standard, these 3" discharges only have to supply 375 gpm and 4" discharges 625 gpm. These values are significantly lower than what 4" or 5" supply lines are capable of flowing.[5]

Although the flows listed in *NFPA 1900* are the minimum requirements, most LDH discharges can flow more than these minimum flow rates. The question then becomes, "How much more?" This is where flow testing individual discharges and determining the maximum realistic flow rate through a particular discharge becomes so important. This determination is done by evaluating the internal plumbing loss characteristics for each discharge at given flow rates.

Once the number and size of the apparatus discharges are recorded, testing to determine the friction loss in the apparatus plumbing can begin. A large diameter supply line should be connected between the hydrant and an intake on the apparatus. For 2½" discharges, a portable monitor should be set up and have one line connected to its inlet. Large diameter discharge testing will require a large diameter supply line to be connected to the discharge and a manifold to be attached to the end of the line. From the manifold, multiple lines can be run to feed one or two portable monitors based on the flow rate to be tested (fig. 2–4). A flowmeter can be connected on the inlet side of the monitor or on the discharge of the monitor depending on the type and style of flowmeter being utilized. A smoothbore master stream nozzle should be attached to the monitor to ensure that the readings on the flowmeter can be verified by a pitot gauge.

The tests should first begin with flowing the NFPA minimum flow rate requirement for each 2½" and larger discharge on the apparatus. Once the

Figure 2–4. Flow test of a large diameter discharge using a single 5" line to a manifold, which is feeding two portable monitors

flow rate is reached and is verified by a pitot gauge, the operator should record the reading on the master discharge gauge. Next, the reading on the individual line gauge of the discharge being tested should be recorded. The internal plumbing loss for the particular discharge being tested can be determined by subtracting the individual discharge reading from the master discharge reading.

After the plumbing loss values for the NFPA minimum flow rates have been determined, operators can determine what the realistic maximum flow rate for each individual discharge is. The amount of internal plumbing loss for a particular discharge at a specific flow rate will dictate the maximum realistic flow rate for the discharge. It is important to remember that when the flow rate through a discharge doubles, the plumbing loss will quadruple. For example, if the plumbing loss through a 2½" discharge is 5 psi at a flow rate of 250 gpm, the loss at 500 gpm will be roughly 20 psi.

It is worth noting that depending on the gauge graduations of each discharge gauge and their accuracy, the friction loss values may not quadruple when the flow rate is doubled. This is especially true at lower flow rates through each discharge. The exponential increase in internal plumbing loss becomes much more apparent at higher flow rates through individual discharges, however.

Fire departments must identify what they feel is an acceptable internal plumbing loss value for a given flow rate. For higher flow rates, fire departments should tolerate a lower internal loss value. This is primarily because high flow rates require higher engine revolutions per minute in order for the pump to move the desired volume of water and build the necessary pressure. If the pump can flow 2,000 gpm through a single discharge, but the internal appliance loss is 100 psi, this means that the pump will have to build an additional 100 psi on top of the pressure requirements for the supply layout being fed. The apparatus could run out of horsepower and reach the governed speed of the motor before reaching the pressure requirement for the given flow rate.

The operator will notice that as the flow rate is increased through a particular discharge, the internal friction loss value will also increase. The increase in the internal plumbing loss will dictate what the realistic maximum flow rate is based on what the department determines to be an acceptable amount of plumbing loss. Table 2–2 shows an example of the information collected by the author during a maximum flow test for a large diameter discharge with 3" plumbing.

A flow rate of 400 gpm was tested first instead of the NFPA minimum of 375 gpm due to the fact that a 1¼" smoothbore tip was used for the initial test. This made for an easy flow verification with a pitot gauge. On this particular rig, the scale of each of the discharge gauges read from 0 psi to 600 psi. This scale made it difficult to accurately pinpoint the internal friction loss values at lower flow rates because the gauge graduations were in 10 psi increments. As a result, the increase in internal appliance loss values from 400 gpm to 625 gpm do not quite follow the hydraulic rule that states friction loss varies with the square of the increase in flow rate. This hydraulic rule is more noticeable once higher flow rates were tested, however.

A review of table 2–2 reveals that the pumper tested could easily flow the *NFPA 1900* minimum flow rate through the large diameter discharge

Table 2–2. Maximum flow rate tests for 3" large diameter discharge

Flow rate (gpm)	Master discharge reading (psi)	Individual discharge reading (psi)	Internal appliance loss (psi)
400	90	80	10
625	115	100	15
1,000	130	100	30
1,500	190	120	70
2,000	260	130	130

with a negligible amount of internal loss. The information in this table also shows what the realistic maximum flow rate is for this particular large diameter discharge. Although the large diameter discharge was capable of flowing 2,000 gpm, it resulted in 130 psi of internal plumbing loss. The absurdly high loss values for this 3" LDH discharge make it unrealistic for this single discharge to reliably support a 2,000 gpm flow. While performing the 2,000 gpm test, the apparatus involved reached the governed speed of its motor.

The flow tests performed in table 2–2 show that for this particular rig, a realistic maximum flow rate for the larger diameter discharge is 1,000 gpm. At this flow rate, the internal appliance loss is only 30 psi for the large diameter discharge. This should be considered an acceptable amount of loss during high flow operations. In addition, when maximum flow tests were performed on the 2½" discharges for this apparatus, it was found that each 2½" discharge was capable of flowing 500 gpm with roughly 30 psi of internal loss as well. The goal should be for the fire department to perform this process for every apparatus in their fleet. This is mainly because different apparatus may have different plumbing configurations that will affect the maximum flow rates through particular discharges.

After testing is completed, pump charts should be developed that contain the maximum realistic flow rates and internal loss values for each discharge on the apparatus. This information will help the operator determine how many discharges should be utilized during high flow operations to achieve a particular flow rate. Using table 2–3 as an example, it could be determined that Engine 1 would have to use its large diameter discharge and at least two 2½" discharges to support a 2,000 gpm supply operation efficiently. After determining the number of discharges that must be used,

Table 2–3. Engine 1 maximum discharge flows and internal plumbing losses

Discharge	Plumbing size (in.)	Maximum flow rate (gpm)	Internal plumbing ILoss (psi)
No. 1 driver's side discharge	2½	500	25
No. 3 driver's side discharge	2½	500	20
No. 2 passenger's side discharge	2½	500	25
No. 4 passenger's side discharge	2½	500	30
Driver's rear discharge	2½	500	40
Large diameter discharge	3	1,000	30

the operator must consider the friction loss characteristics in the supply hose connected to each of these discharges. Hose configurations attached to each of these discharges will be discussed in further detail in later chapters.

Study Questions

1. Smoothbore master stream nozzles are traditionally pumped at what nozzle pressure?

   a. 50 psi
   b. 80 psi
   c. 100 psi
   d. 160 psi

2. When pumping to another engine, what must the operator be sure to account for in the calculations?

   a. Nozzle pressure
   b. Residual intake pressure
   c. Residual discharge pressure
   d. Net pump discharge pressure

3. Which of the following has the biggest impact on friction loss?

   a. Hose length
   b. Elevation
   c. Flow rate
   d. Residual intake pressure

4. What is the friction loss in 900' of 4" hose flowing 1,000 gpm?

   a. 20 psi
   b. 100 psi
   c. 120 psi
   d. 180 psi

5. When determining the pump discharge pressure for a supply operation, the operator should account for at least 20 psi of residual intake pressure on the next engine, but no more than ____________.

   a. 30 psi
   b. 50 psi

c. 100 psi
d. 150 psi

6. Which of the following is the condensed Q formula for 5" hose?

a. Q^2
b. $Q^2/5$
c. $Q^2/15$
d. $Q^2/20$

7. Which of the following is an alternative method for determining the friction loss in 5" hose on the fireground?

a. The hand method
b. The drop ten method
c. The rule of 12s
d. The double drop zero method

8. Municipal pumpers typically have one large diameter discharge that does not exceed:

a. 2½"
b. 4"
c. 6"
d. 10"

9. How should a fire department determine the maximum realistic flow rate for each discharge on a pumper?

a. Ask the apparatus manufacturer.
b. Base it on the values listed in *NFPA 1900*.
c. Flow test the apparatus.
d. Do not worry about it.

10. A supply pumper is feeding an attack engine through 700' of 4" hose. The supply pumper is flowing 1,000 gpm through the supply line, and there is no change in elevation. The supply engine should pump at ______________ in order to appropriately supply this volume to the attack engine.

a. 50 psi
b. 100 psi
c. 140 psi
d. 170 psi

3

PRINCIPLES OF SUPPLY OPERATIONS

Serving as the operator of the supply pumper is often a firefighter's last choice. This is primarily because there is a stigma in the fire service that water supply operations are not as "sexy" as attack engine operations. After all, the attack engine operates close to the fire scene and has a direct, tangible effect on the firefighting operation. Firefighters often forget that the role of the supply pumper is quite possibly one of the most important roles on the fireground. Without a competent supply pump operator, the availability of water at the fire scene will suffer dramatically.

This chapter will lay the groundwork regarding the major concepts of supply pumper operations. The remainder of the text will discuss the options available to pump operators for maximizing their available flow rates under various circumstances. Operators must remember that it is their responsibility to overcome any obstacle that interferes with the establishment of a reliable water supply and to provide the greatest amount of water from that source. Although the role of the supply pump operator is not the most glamorous one, it has an enormous impact on the success of the entire operation.

Supply operations are often simply summarized with the statement, "Get water and send it to the fire scene." While this statement is true, the operator must account for several other variables to successfully supply water to the scene. Operators must remember that just getting water to the fire scene is not enough. If this were truly the case, it would be acceptable to forward lay 1,000' of 2½" hose from a fire hydrant, hook the line to an outlet on the hydrant, and then simply use hydrant pressure to feed the attack pumper. While this does "get water to the fire scene," the flow rate available to the attack pumper would be pitiful.

The more appropriate mindset that pump operators should adopt regarding water supply operations is to maximize the availability of water to the fire scene. In the example above, this can be accomplished by making two major changes to the operation. First, large diameter hose (LDH) should be utilized instead of 2½" hose for the supply line. Second, a supply engine should be assigned to pump the hydrant. These two changes in this hypothetical scenario will drastically increase the amount of water available to the attack engine at the fire scene.

In order to maximize the water available to the fire scene, the pump operator must understand the variables that affect the volumetric flow rate from a water source. These variables can be summarized using the following equation for volumetric flow rate *(Q)*, which is referred to as the *continuity equation:*[1]

$$Q = A \times V$$

where

A = Cross-sectional area of the opening
V = Velocity of the water

This formula shows that the flow rate through a system can be increased by increasing either the area of the opening or the velocity of the water, or both. A water source for a firefighting operation is usually either a fire hydrant or a static water supply of some sort. When we evaluate the continuity equation, we see that the pump operator can realistically influence only one of the variables in the equation to increase the flow rate from the water source, which is the area of the opening.

When operating from a fire hydrant, this increase in cross-sectional area is achieved by performing a heavy hydrant hookup (fig. 3–1). During drafting operations, it is accomplished by setting up either a twin or triple tube drafting configuration (fig. 3–2). The key here is to understand that the only way to maximize the available water from a water source is to utilize multiple intakes from that source to increase the total opening area feeding the pump. Doing so splits the total flow rate entering the pump between multiple intakes, which in turn reduces the total friction loss in the entire supply side of the system. The result will be higher fire flows during both hydrant and drafting operations.

Figure 3–1. Heavy hydrant hookups are performed in order to increase the area feeding the pump.

Figure 3–2. Drafting through multiple intakes increases the area feeding the pump when operating from a static source.

The pump operator does not have the ability to directly increase the velocity of the water from the source when the outlet area remains constant. Doing so from a fire hydrant would require the fire hydrant to have a built-in pump that could increase the velocity of the water when the outlet area remained the same. The only way to increase the velocity of the water when operating from a draft would be to somehow increase the atmospheric pressure acting on the water to cause the water to move faster through the hard sleeve. Neither one of these scenarios is realistic, meaning that the only true option available for increasing the available flow from a water source is to increase the total area feeding the supply pumper.

It should be noted that by increasing the area of the opening, the operator does indirectly affect the velocity of the water. If the flow rate remains the same after the area feeding the pump is increased, then the velocity of the water decreases. This is because the water does not have to move as fast to deliver the same volume through a smaller opening.

Utilizing the continuity equation along with the friction loss principles discussed in chapter 2, it becomes apparent that utilizing multiple intakes with large supply hose will allow higher flow rates into the pumper and will reduce the overall friction loss within the supply system. The goal of any supply operation should be to get as much water into the supply pumper as possible. From there, the supply pump operator has control of it. The pump operator should consider this concept to be the backbone of supply pumper operations.

Positioning the Supply Pumper

The effective establishment of a water supply requires the operator to first position the apparatus properly. When positioning the supply pumper, the operator has more than one consideration. The pump operator must be mindful not only of getting water into the pumper, but also of allowing access for other apparatus. In urban or suburban settings, this typically means leaving room for a truck company to access the scene. During rural water supply operations, this means leaving enough room for water shuttle apparatus to access the dump and fill sites. Failure to consider the needs of other apparatus on the scene can have a significant negative effect on the operation as a whole.

When positioning the supply pumper, the operator should also be mindful of any intakes that will be utilized for the operation. As discussed in chapter 1, pump intakes come with their own sets of advantages and disadvantages. One of the major advantages of side intakes is that they are usually the best-plumbed intakes on the apparatus. This means that they will allow the most amount of water into the supply pumper with a minimal amount of loss (fig. 3–3). The major disadvantage of these intakes is that positioning the apparatus for their use can be difficult, especially on narrower streets. If the apparatus is too close to the source, it is likely that a kink will form in the supply line (fig. 3–4). Positioning further from the source may cause the apparatus to be too far in the roadway, which could block access for other apparatus arriving on scene (fig. 3–5).

Figure 3–3. Side intakes feed water directly into the pump with minimal plumbing loss.

Figure 3–4. Kinks in the supply line will likely result if the apparatus is positioned too close to a fire hydrant and a side intake is utilized. This condition is exacerbated if using pony lengths 25' or shorter.

Figure 3–5. Positioning further from the hydrant prevents a kink from forming, but it blocks other apparatus from being able to access the scene.

In contrast, use of front intakes typically includes more plumbing losses, but their location on the pumper allows the apparatus to be positioned in a more advantageous fashion. When dealing with a hydrant, the operator can simply hug the curb and nose in toward the hydrant. This allows the supply pumper to leave room for other arriving apparatus (fig. 3–6). The other major advantage to positioning in this fashion is that the pump operators have the ability to see the intake and the water source the entire time they are spotting the apparatus. This will allow the operators to position the apparatus close enough to reach the water source, but not so close that a kink in the supply line will form.

When operating as a supply pumper at a dump site in the rural setting, it is rarely possible to set up a dump tank where the operator can draft from a side intake. Doing so would be likely to block access for tankers coming to off-load their water. For these reasons, it is typically advisable to set the dump tank either in front of the pumper or behind the pumper to leave at least one lane of travel open for the tankers to off-load (fig. 3–7). This means the supply pumper will either back or pull into a driveway or stay in one lane of travel in the street. Having a front intake and/or a rear intake in these situations allows the apparatus to position with functionality at the forefront.

Nosing in toward a water source is also advisable in certain situations when operating as a fill site engine in the rural environment. This could

Figure 3–6. By using the front intake on this apparatus, the operator is able to rapidly access water from the hydrant while also keeping a lane open for other apparatus.

Figure 3–7. Dump tanks should be set up either in front of or behind the dump site pumper in order to maximize the space available for mobile water supply apparatus. This is where having a front intake can become extremely valuable to the operator.

simply be because positioning sideways for a side intake would require additional hard sleeve to reach the water source compared to nosing in toward the source (fig. 3–8). There may be other times in the rural environment where the apparatus may be forced to leave the pavement in order to reach the water source for a fill operation. Pump operators must understand that this is a highly risky operation, and they must always thoroughly assess the surface before taking the apparatus off the pavement. If the pump operator walks the surface and notes that the ground is soft underfoot, under no circumstances should the apparatus be taken off the pavement. If it is determined that the surface cannot support the weight of the apparatus, the operator must find a different water supply solution.

Positioning with the nose of the apparatus toward the water source becomes even more important when the apparatus is positioned on an unpaved surface. If the apparatus does become stuck after committing off the pavement, this orientation of the rig will make it easier for a wrecker to pull the apparatus out (fig. 3–9). Operators who attempt to position the apparatus sideways alongside the water source when operating off the pavement typically loosen the ground in the process. The result is an apparatus that is stuck sideways, which will require a much more complex recovery operation. Although front intakes will make it easy to establish a water supply after the rig has been positioned in this fashion, the technique of positioning with the nose in should be performed regardless of the presence of a front intake.

Figure 3–8. Nosing in toward this source allowed this fill site pumper to rapidly establish water with a minimal amount of hard sleeve. As equipment becomes available, the operator will utilize multiple intakes to increase the fill rate.

The goal of every supply operation is to establish a reliable water supply and maximize the water availability to the fire scene. As stated earlier, this is accomplished by utilizing multiple intakes simultaneously in order to increase the total area feeding the pump. When positioning the supply pumper, pump operators must take this into consideration. Not only must they be concerned about the initial intake utilized to establish the water supply, but they must remember to position the apparatus so that additional intakes can be utilized to maximize the water availability (fig. 3–10).

Figure 3–9. By nosing in toward this source, this operator has made it easier for the wrecker to pull the apparatus out in the event it becomes stuck.

Figure 3–10. The operator should position the apparatus so that multiple intakes can be utilized simultaneously. This will increase the available water to the fire scene.

The Fast Water Process for Initial Operations

The supply line feeding the initial attack engine is typically laid using one of three deployment methods: a *forward lay* performed by the attack engine, a *reverse lay* performed by the supply pumper, or a *booster backup operation.* In each scenario, the supply pump operator must understand that the goal is to give the attack engine uninterrupted water as quickly as possible. The first step in determining which deployment method to utilize is to identify what the primary water supply is and determine where it is located.

Once the primary water source has been identified by the first-due engine company, the location and supply line deployment tactic should be communicated to the other responding units. Communicating the water supply plan is vital to ensure the supply pumper is able to rapidly establish water to the attack engine. After appropriately positioning the supply pumper, the operator is ready to work on the establishment of a water supply. Just as the initial attack engine uses their tank water for the initial fire attack, the supply engine should be prepared to utilize their tank water for the initial supply. This is known as the *fast water process.*

Fast water for forward lay scenarios

If the attack engine elects to perform a forward lay, the supply pumper will be responsible for positioning at the water source and establishing the water supply. In scenarios involving a hydrant, the attack engine will wrap the hydrant and then perform the forward lay. In the rural environment, forward lay operations typically occur when the attack engine is laying up the driveway to access the fire scene. In either case, it will be the responsibility of the supply engine to position where the attack engine began the forward lay and then begin establishing the water supply.

After the supply engine has been positioned appropriately, the next step is for the operator to engage the pump of the supply engine and recirculate water. The operator should then connect the supply line laid by the attack engine to a discharge on the supply pumper. At this point, the operator must make a critical radio transmission to the attack engine stating that the supply engine is ready to send them the water in their booster tank. This transmission lets the attack engine know that the initial water coming is limited.

The attack engine will reply back to the supply engine stating that they are either ready for the supply pumper to send their tank water or that they are not. This will be dependent on the fire conditions at the scene and the arrival times between the initial attack engine and the supply pumper. If the attack engine tells the supply engine to send their tank water, then the operator of the supply engine should open their discharge and begin doing so. Once the discharge is opened and tank water begins flowing to the attack engine, the supply operator should begin establishing the external water supply. This means making hydrant connections to the supply pumper when dealing with a fire hydrant or setting up a dump site in the rural environment.

In the event that the attack engine radios to the supply pumper and states that they are not ready for the supply line to be charged, the supply operator should start making connections from the water source to the supply pumper. When this occurs, it may be likely that the supply engine is able to establish a water supply before the attack engine calls for water. If at any point during the establishment of the water supply the attack engine calls for water, however, the supply pumper can simply open their discharge and start sending tank water to the attack engine while they finish establishing the water supply. Once the external water supply has been established, the supply pumper must radio the attack engine to inform them of this tactical benchmark.

When operating from a fire hydrant, the fast water process allows the attack engine to have some additional water should they need it while the supply pumper hooks to the hydrant. In addition, it also puts the supply engine in a position to be able to pump the hydrant and overcome the friction loss in the supply line. This will allow more of the hydrant's capacity to be delivered to the fire scene, especially on longer layouts. This configuration also adds a layer of redundancy to the supply operation. Should the attack engine have a mechanical failure, the supply engine is tied into the system and can easily pump the fire from the hydrant.

Certain departments may be leery of dedicating a pumper to the hydrant on every fire, especially since many fires do not require the full capacity of the hydrant being utilized. An option for these departments is to use a four-way hydrant valve instead. These valves allow the attack engine to perform a forward lay and initially operate off of hydrant pressure alone. If the fire flow demands increase later in the incident, however, a pumper can be assigned to hook to the four-way valve and boost the pressure to the attack engine without having to shut down the hydrant (fig. 3–11). Four-way hydrant valve operations will be discussed in greater detail in chapter 7.

Figure 3–11. Four-way hydrant valves are extremely useful for departments that routinely perform forward lays.

Fast water for reverse lay scenarios

Reverse lay operations are quite similar to forward lay operations from the standpoint of the supply pumper. The major difference between the two is the apparatus that lays the supply line. During a reverse lay scenario, the attack engine will go directly to the fire scene without laying a supply line. The supply pumper will be responsible for dropping their supply line at the attack engine and then driving toward the water source. Once the supply engine positions appropriately at the water source, the supply line must be broken from the hose bed and connected to a discharge in the pumper.

At this point, the operator should radio the attack engine to inform them that they are ready to send their tank water. Just as outlined above, the decision to send tank water from the supply engine will be dependent on whether the attack engine immediately needs it. If the attack engine calls for the supply engine's tank water, the supply engine should send it before working on the external water supply. If the supply engine is told to wait, then the supply operator can begin to hook up the supply pumper to the water source. Once the supply pumper is hooked up to the external water source, the operator should radio the attack engine to inform them that a reliable water source has been established.

Booster tank size and LDH fill capacities

The operator must remember that it takes water to fill and charge any hose line. The larger the hose line, the more water it takes to charge it. This can become critically important when determining if tank water should be used to initially fill the supply line. After all, the amount of tank water that will actually be available for the attack engine to use will be dependent on the length of the hose lay, the size of the hose, and the capacity of the supply engine's booster tank. Table 3–1 compares the amount of water it takes to fill various supply lines based on the length of the supply layout.

Using the information provided in table 3–1, we can determine how much water would be available to the attack pumper if a supply engine with a 750-gallon booster tank were to charge 1,000' of supply line. If a 3" supply line was laid, the attack engine would be able to use 380 gallons of water from the supply engine after the line was charged. A 4" supply line would allow the attack engine to use 100 gallons of the 750 gallons in the supply engine's booster tank. If 5" hose was laid, the water in the booster tank would not even be sufficient to completely charge the 1,000' of supply hose.

In the scenario above using 5" hose, the question becomes whether or not the supply engine should send their tank water if they know it will not even reach the attack engine. The recommendation is for the supply engine to send their tank water if the attack engine calls for it, even if all of the tank water is used just to fill the supply line. Water will always be used to initially fill the supply line. It does not matter if that water comes from the supply engine's booster tank, the fire hydrant, or a dump tank; the debit must be paid. If the attack engine calls for the supply engine's tank water, the supply operator should radio and inform the attack engine that tank water is on the way, but that it may not reach the attack engine. After the water is sent, the supply operator must rapidly work to finish establishing the external water supply. This will ultimately lead to a faster establishment of water for the attack engine. By filling the supply line with tank water

Table 3–1. Volume of water required to fill various hose layouts

Hose size	Gallons used to charge 100'	Gallons used to charge 500'	Gallons used to charge 1,000'
3"	37	185	370
4"	65	325	650
5"	100	500	1,000

Note: Values are rounded to the nearest whole number.

while simultaneously establishing an external water supply, the supply operator is essentially killing two birds with one stone.

Dealing with excessive intake pressures on the attack engine

One of the major disadvantages to tying the supply pumper into the hydrant from the beginning of the operation is that there is the potential for the attack engine to be overpumped. This typically occurs when the hydrant has a high residual pressure and the flow rate at the fire scene is low. The operator must remember that when the supply engine's pump is engaged, even with the motor at idle, pressure is still built on the discharge side of the pump. This pressure is typically anywhere from 30 psi to 50 psi. This means that if the residual intake pressure from the hydrant is 70 psi, the pressure leaving the supply pumper at idle is around 120 psi. At low flow rates through 4" or 5" supply lines, the attack engine would have to deal with excessively high intake pressures coming into their pump.

If the attack engine is reporting excessively high intake pressures even with the supply engine at idle, the operator of the supply pumper should put the vehicle transmission in neutral (N). This will stop the supply pumper from building pressure on the discharge side of the pump and will allow the water from the hydrant to simply flow through the supply engine. At this point, the attack engine is being fed solely by hydrant pressure, and they should see a reduction in pressure on their intake gauge. With the transmission of the supply pumper in neutral, the supply engine becomes nothing more than a giant manifold.

The operator of the supply pumper only needs to shift the vehicle transmission to neutral and should not move the pump shift lever from the PUMP position to the ROAD position. With water flowing through the plumbing of the pump, there is the potential for the water flow to cause the impeller to spin slightly. This in turn will cause the gears and/or chain drive in the transfer case to turn as well, which will result in the grinding of gears if the pump shift lever is manipulated. Simply putting the vehicle transmission in neutral is sufficient and will not cause harm to the apparatus. In the event that the fire flow demands increase, the operator of the supply engine can easily put the vehicle transmission back in drive (D) and begin pumping again.

The supply pumper and booster backup operations

Some fire departments elect to use a tactic known as *booster backup operations* for initially establishing water to the attack engine. Using this tactic, the first-arriving engine proceeds directly to the scene and initiates a fire attack utilizing tank water. When the second pumper arrives, they are responsible for stopping at the water source, dropping their supply line, and performing a forward lay to the fire scene. In the urban/suburban environment, this involves wrapping a hydrant and performing a forward lay to the fire scene. Rural scenarios typically involve the supply pumper dropping the supply line at the mouth of the driveway and then forward laying up to the fire scene.

After performing the forward lay, the second-arriving pumper must position appropriately and engage their pump. The whole intention of the booster backup operation is for the second-arriving pumper to give their tank water to the initial attack engine. This expands the length of time that the initial attack line can operate while a sustainable water supply is being established. For example, if both the initial attack engine and the second-due pumper are equipped with 750-gallon water tanks, a booster backup operation will allow for just under 1,500 gallons of water to be utilized for the initial fire attack while the water supply is secured.

The exact amount of water available to the initial attack engine through a booster backup operation will vary depending on the diameter and length of the supply line used to feed the attack engine. In light of this, operators of the second pumper should only run 2½" or 3" hose from a discharge on their engine to the auxiliary intake of the attack pumper. The major reason for this is that it requires less water to charge 2½" or 3" hose compared to charging 4" or 5" LDH. This allows more of the water from the second pumper to be utilized for firefighting operations as opposed to being used to charge the supply line. In addition, 2½" or 3" hose will be much easier for the operator to deploy between the two rigs compared to LDH supply line.

Some operators are hesitant to utilize medium diameter hose such as 2½" or 3" hose for the booster backup operation because they are concerned about flow restrictions. The pump operator must remember that the booster backup operation is only intended to expand the operating period of the initial attack line by giving the second engine's tank water to the attack pumper. As discussed in chapter 1, *NFPA 1900* requires that the tank-to-pump flow rate on apparatus equipped with 500 gallon or larger tanks be a minimum of 500 gpm.[2] Unless the department specifies a higher tank-to-pump flow rate, 500 gpm is what most pumpers are capable of moving through their tank-to-pump line. This means that for most apparatus, the maximum flow rate of the initial attack engine operating off their booster tank will be 500 gpm. Since the second-arriving pumper is also operating off of their booster tank during a booster backup operation, the maximum flow rate that they will be able to supply the attack engine with is also 500 gpm.

The operator must remember that 2½" or 3" hose will be more than capable of supplying 500 gpm. Ideally, the second pumper is capable of positioning within 100' of the initial attack engine when employing the booster backup. This means that if 2½" hose is being utilized for the booster backup operation between the two rigs, there will be 50 psi of friction loss in the 100' of 2½" hose at 500 gpm. If the pumpers are equipped with 3" hose instead, the friction loss in 100' of hose would only be 20 psi at 500 gpm. In both of these scenarios, the utilization of medium diameter hose allows for the rapid establishment of the booster backup operation while maintaining an adequate flow rate for the attack crews (fig. 3–12).

Once the operator of the second pumper has begun feeding the tank water on the apparatus to the attack engine, the next step is to work toward establishing a sustainable water supply. This begins by determining which pumper to hook the supply line to. The operator should go to the rear of the second pumper and disconnect the supply hose that was previously laid. The supply hose should then be connected to an intake on the initial attack pumper. Once the supply line is charged, the attack engine will have a reliable source of water for firefighting operations.

Figure 3–12. After laying in from the water source, the second engine should initiate a booster backup operation by running medium diameter hose from a discharge on their pumper to the auxiliary intake of the attack engine.

Many operators want to take the supply line and instead connect it to an intake on the second pumper. This choice creates the potential for excessive intake pressures on the attack engine because the two pumpers are operating in a series. While this is not the worst thing to happen on the fireground, it makes it more challenging for operators of the attack engine to appropriately manage their flow pressures. The better option is to hook the supply line directly to the initial attack engine because it is ultimately the one that needs the water.

After the supply line has been connected to the attack pumper, water can be sent from the water source. When operating from a hydrant, this may involve hooking the supply line directly to the hydrant and turning the hydrant on. While this is a less-than-ideal water supply configuration, it is typically done when the third-due engine has a longer response time or may be delayed. In these scenarios, it is extremely advantageous to attach a four-way hydrant valve to the *steamer port*, which is typically a 4½" hydrant connection, prior to opening the hydrant. This will allow a later-arriving pumper to hook to the four-way valve and boost the pressure in the event that it is required.

The preferred option when utilizing a fire hydrant as a water source is to assign the third-arriving engine to pump the hydrant and serve as the supply engine. This will allow the third pumper to overcome the friction loss in the supply line and provide more of the hydrant's capacity at the fire scene should it be needed. When using the booster backup in the rural environment, a third pumper will be positioned at the mouth of the driveway to supply water to the scene. This third engine will serve as the dump site pumper. Dump site operations will be discussed in more detail in chapter 7. In both of these scenarios, the operator of the third engine will be responsible for making the necessary connections and sending water to the scene. At this point in the operation, the attack engine should have a reliable external water supply established.

Relay Pumping Operations

Relay pumping operations are one of the most common tactics utilized to supply water to a fire scene. Most firefighters think of this as a tactic for supplying water over long distances. Regardless of the distance, the simplest relay pumping configuration utilizes two pumping apparatus: an attack pumper and a supply pumper. Certain scenarios will require that relay pumpers be placed in between the supply and attack pumpers. This allows higher volumes of water to be delivered to the fire scene by reducing the distance between each pumper and therefore reducing the friction loss.

Since only two pumpers are required for a relay pumping operation to take place, each of the fast water scenarios discussed above are forms of relay pumping operations. Relay pumping operations are viable tactics to utilize in both the urban/suburban and rural environments. There are three major variables that dictate how many pumpers are needed to effectively perform a relay operation: the size of the supply line utilized, the distance between the scene and the water source, and the volume of water required at the scene.

Supply line diameter

The diameter of the supply line utilized in the relay pumping operation has a significant impact on the number of pumpers needed, the maximum distance between each pumper, and the total volume that can be supplied. As discussed, the smaller the supply line, the greater the friction loss. Since larger supply lines have lower friction loss values, pumpers can be spaced further apart depending on the flow rate being supplied through the relay. This is advantageous for organizations with limited resources. The larger the supply line, however, the more water it will take to fill. This may not be an issue in the urban or suburban environment, but in the rural environment, this could be a huge problem. Many rural companies elect to utilize 3" hose as their supply line because it requires less water to fill and allows more water to be available at the scene for firefighting operations.

It is important to understand that if medium diameter hose is utilized for any supply operation, multiple supply lines must be laid. This allows a sufficient volume of water to be supplied to the scene. Many rural companies that are running 3" hose as their supply line are doing so by utilizing a *split hose bed* (fig. 3–13). One of the advantages of this hose bed configuration is that it allows for the deployment of dual lines any time a forward or reverse hose layout is performed. The major advantage to laying dual 3" lines is that it allows for only one of the supply lines to be charged initially to feed the fire scene. This permits rural companies to conserve water until a reliable supply is established. In the event more water is required by the attack engine, the second 3" supply line can be charged. Hydraulically speaking, dual 3" lines are capable of moving the same volume of water as a single

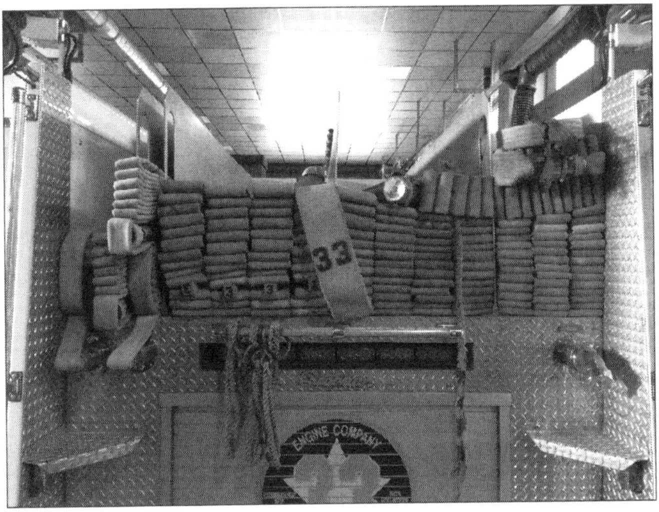

Figure 3–13. This engine is configured with a split bed of 3" hose. This allows the company to lay dual supply lines that hydraulically equate to a single 4" supply line.

4" supply line with the same friction loss values. This gives rural fire companies the flow advantages of LDH without having to switch to it completely.

Table 3–2 outlines the number of medium diameter hose (MDH) supply lines needed to move the same volume of water as a single large diameter supply line. This table also outlines the friction loss characteristics for the single LDH supply line as well as the friction loss characteristics in each MDH supply line required for the equivalency. It should be noted that when an MDH supply line is utilized for supply operations, it is advantageous to

Table 3–2. Required number of MDH lines to flow the same volume as a single LDH line

Medium diameter hose layout	Large diameter equivalent	Friction loss per 100' at 1,000 gpm in single LDH (psi)	Friction loss per 100' in each MDH (psi)	Flow rate in each MDH to supply 1,000 gpm total (gpm)
Dual 3" Lines	Single 4"	20	20	500
Three 2½" Lines	Single 4"	20	~22	333
Three 3" Lines	Single 5"	8	~9	333
Five 2½" Lines	Single 5"	8	8	200

Note: Friction loss values listed as "~" are rounded to the nearest whole number. The difference should be considered negligible.

utilize 3" hose rather than 2½" hose for the operation. This is primarily because less 3" hose has to be put on the ground compared to 2½" hose to gain the same flow characteristics as a single large diameter supply line.

Distance and flow rate

The distance between the fire scene and the water source has a significant impact on the design of the relay operation. Coupled with the hose diameter utilized, the distance will affect the number of pumpers required for the operation. Many firefighters believe that the distance of the layout has the biggest impact on the design of a relay pumping operation. While it is a key factor, in reality, the flow rate through the system has the biggest impact on how the relay operation must be designed. As discussed in chapter 2, this is due to the fact that when the flow rate through a hose line is doubled, the friction loss is quadrupled.

Table 3–3 outlines how far pumpers can be spaced from each other for a given flow rate through the supply line and at a specific pump discharge pressure. Obviously, the lower the flow rate through the supply line and the larger the hose diameter, the further each pumper can be spaced apart in the relay operation. As the flow rate through the supply line increases, the distance between each pumper must decrease. This will require more pumpers in the relay to provide higher flow rates over longer distances.

The pump operator must be cognizant of the pump discharge pressure feeding the supply line. One reason is to ensure that the service pressure of the supply hose is not exceeded. In most instances, especially when LDH is used, the weakest link in the supply operation is the supply hose itself. According to *NFPA 1961, Standard on Fire Hose*, supply hose 3½" and greater must be capable of withstanding a minimum service test pressure of 200 psi.[3] Some hose manufacturers produce supply line capable of withstanding 250 psi or 300 psi service test pressures, but the minimum requirement is only 200 psi. Therefore, an operator who is unsure of the rating of the supply hose in use should assume that it only has a service test pressure of 200 psi. In addition, *NFPA 1962: Standard for the Care, Use, Inspection, Service Testing, and Replacement of Fire Hose, Couplings, Nozzles, and Fire Hose Appliances* states that hose marked SUPPLY HOSE should not be operated at pressures that exceed 185 psi.[4] This 15 psi difference between the service test pressure and the recommended maximum operating pressure serves as a safety factor to help prevent failure of the supply line during fireground operations.

Table 3–3. Maximum relay distances with varying hose diameters

Flow rate (gpm)	Friction loss (psi/100')	Relay distance at 150 psi PDP (ft)	Relay distance at 185 psi PDP (ft)	Relay distance at 200 psi PDP (ft)
Maximum relay distance through single 3" LDH				
250	5	2,600	3,300	3,600
500	20	650	825	900
750	45	275	350	400
1,000	80	150	200	225
Maximum relay distance through single 4" LDH				
500	5	2,600	3,300	3,600
750	11	1,150	1,500	1,600
1,000	20	650	825	900
1,500	45	275	350	400
2,000	80	150	200	225
Maximum relay distance through single 5" LDH				
500	2	6,500	8,250	9,000
750	4.5	2,850	3,650	4,000
1,000	8	1,625	2,050	2,250
1,500	18	700	900	1,000
2,000	32	400	500	550

Note: Calculations include 20 psi residual intake pressure for the next engine in the relay. PDP is pump discharge pressure.

It is also important to consider the effects that higher pump discharge pressures have on the rated capacity of a pump. As discussed in chapter 1, centrifugal fire pumps are rated to flow their maximum rated pump capacity at 150 psi, 70% of their capacity at 200 psi, and 50% of their capacity at 250 psi while drafting. Even if a department is using large diameter supply hose that has a higher service test pressure rating, it is not advisable to exceed a net pump discharge pressure of 200 psi during a drafting operation. This is because net pump discharge pressures greater than 200 psi will result in the pumper being only capable of reliably supplying 70% or less of its rated pump capacity.

Net pump discharge pressure or *net PDP* is defined as the amount of pressure that is actually created by the pump, which is essentially how much work the pump is doing. When operating from a draft, the pumper not only has to create a vacuum at the eye of the impeller to draw water into the pump, but it also has to build pressure on the discharge side of the pump. When the pumper is operating from a pressurized source, the pump takes advantage of the pressurized water that enters from the outside source. This basically means that the pump does not have to work as hard to build pressure when operating from a hydrant as when operating from a draft.

Table 3–3 has the maximum relay distances broken down by three pump discharge pressure categories and assumes that the pumper is drafting. This is considered the worst-case scenario because the pumper receives no positive pressure assistance. The 150 psi column represents the furthest distance that pumpers may be spaced while maintaining 100% of the rated capacity of the pump. The 185 psi column outlines the maximum allowable distances between pumpers without exceeding the *NFPA 1962* pressure recommendation previously discussed.[5] The 200 psi column lists the maximum allowable distances at given flow rates for departments that are using supply hose with higher service test pressure ratings. The operator should also note that the pump pressures listed in table 3–3 include 20 psi residual intake pressure on the next pumping apparatus in the relay.

Pump operators often ask, "What is the maximum volume and distance that is achievable through a relay pumping operation?" In order to determine this, pump operators must identify the supply hose being used and evaluate the discharge plumbing on their apparatus. As chapter 2 discussed, the plumbing configurations of most large diameter discharges are capable of reliably supplying 1,000 gpm with a minimal amount of internal friction loss. Flow rates above 1,000 gpm start to cause excessive plumbing loss within the pump, however. For these reasons, it is generally acceptable to consider 1,000 gpm as the maximum reliable flow rate through a single large diameter discharge for a relay operation. It is important to note that some apparatus may be able to flow more than 1,000 gpm through their large diameter discharge depending on the plumbing configuration of the apparatus. The only way this can be determined is by conducting a flow test of the apparatus plumbing, as outlined in chapter 2.

After the maximum reliable flow rate is established, the operator can determine the realistic maximum relay distance that can be achieved. This determination will be based on the diameter and pressure rating of the supply hose used by the fire department. Table 3–3 can be utilized to determine what the maximum relay distance is for a given flow rate through the supply line carried by the organization. In order to determine how many pumpers are needed for a relay operation, the following formula can be used:[6]

$$\text{Number of Pumpers} = \frac{\text{Total Distance from Source to Scene}}{\text{Max Flow and Distance Value from Table 3-3}} + 1 \text{ Pumper}$$

The reason the "1 Pumper" term is added in this equation is to account for the attack engine located at the fire scene. Since it is not possible to have a fraction of an engine, the final value should be rounded up to the nearest whole number. This formula is useful when preplanning relay operations for specific target hazards. When relay operations are deployed as a necessity for a particular incident scene, the operator should understand that the distances between each engine may not be equal. This is why all pump operators participating in a relay pumping operation must be capable of calculating the necessary pump discharge pressure for the flow rate and hose layout presented to them. The subsections that follow outline the variables that must be taken into account when determining the required pump pressures for relay operations on scene.

Elevation loss in relay operations

Elevation loss is one of the more challenging variables to account for in a relay pumping operation. This is simply because in many scenarios, the pump operator may not know that it exists. Since it is exceedingly rare for two pumpers to see one another directly, it is possible that there could be an elevation rise or fall between them. The amount of elevation change that is present between the two pumpers could have a dramatic impact on the pump discharge pressures required for the operation.

When attempting to compensate for elevation change on the fireground, pump operators have to use their best judgement in order to determine what the difference in elevation actually is. In these scenarios, the operator may undercompensate or overcompensate for the elevation presented to them. If there is more elevation change than the operator calculated for, the next pumper being supplied in the relay may radio and say that they do not have enough incoming intake pressure. In scenarios where there is actually less elevation change than the pump operator accounted for, the receiving engine may radio back and say that they have too much incoming pressure. Both of these scenarios require the pump operator to adjust the pump discharge pressure accordingly.

Certain scenarios may occur in which the increase in elevation is so great that it negatively affects the available flow rate at the scene. In this case, so much pressure must be produced to overcome the elevation loss that the pressure allowance for friction loss and residual intake pressure is diminished (fig. 3–14). This ultimately results in lower flow rates at the fire scene because of concerns regarding the pressure rating of the supply hose. One solution to this problem is to add more relay pumpers to the operation, but this may not be feasible if the operation is already established. A more realistic approach to tackling this problem is to implement a parallel supply line operation. This will allow about twice the amount of water to be pumped up the hill at the same pump discharge pressure.

Not every elevation scenario involves pumping water uphill. Plenty of scenarios exist where a relay pumping operation may be supplying water to a scene downhill from the source. In these situations, gravity will work favorably with regard to building pressure. During a downhill scenario, the pump operator must be mindful not to overly pressurize the intake of the next pumper being supplied in the relay. In order to prevent this, the pump operator may have to idle the throttle back and gate the discharges feeding the next pumper. If this still results in too much intake pressure on the next pumper, the operator should remember the option of placing the vehicle transmission of the pumper in neutral to prevent roughly 50 psi of idle pressure from being added to the water.

Determining the flow rate

In order to appropriately supply water to the fire scene, the supply and relay pumpers must know the required flow rate at the fire scene. It is rarely as

Figure 3–14. Changes in elevation can have a dramatic impact on the total volume of water that can be supplied through a relay operation.

simple as radioing to the attack engine and asking how much water is flowing, however. Even if the supply engine can ask the attack engine their flow rate, the attack engine will usually respond with the number and size of lines they have in operation rather than give a specific flow rate in gallons per minute. For these reasons, the supply and relay operators should mentally keep track of the number and size of lines that are being put in service as the incident progresses. This requires the pump operator to have a thorough knowledge of the flow rates of each line utilized within the department. As a general rule of thumb, table 3–4 outlines average accepted flow rates in the industry for lines of varied sizes.

Using this information, the operator should be able to determine the flow rate at the fire scene fairly accurately. For example, if the attack pumper has two 1¾" handlines in service, the supply engine knows that the fire flow is approximately 300 gpm. If the attack engine reports that they are supplying a deck gun and one 2½" line, the supply engine can safely assume that between 750 gpm and 1,250 gpm will be flowing at the scene. The operator can now use this information to calculate the necessary pump discharge pressure for the operation.

As discussed in chapter 1, the *NFPA 1900* minimum tank-to-pump flow rate for apparatus equipped with 500 gallon or larger booster tanks is 500 gpm.[7] This holds true for all pumpers unless a fire department specified a larger flow rate through the tank-to-pump plumbing when ordering their apparatus. Since this is usually the exception, not the rule, it is safe to assume that most pumpers can flow a maximum of 500 gpm through their tank-to-pump plumbing. This knowledge makes initial supply calculations extremely simple for the supply pumper.

The supply operator should start the necessary pump calculations for the supply operation at an initial flow rate of 500 gpm. This not only makes it easy for the supply operator to calculate the initial pump discharge

Table 3–4. General flow rates

Attack line/Monitor	Flow rate (gpm)
1¾" Hose Line	150
2½" Hose Line	250
Portable Monitor (Blitzfire/RAM)	500
Deck Gun	500–1,000
Aerial Ladder	500–1,000
Elevating Platform	1,000–2,000

pressure for the operation, but it also ensures that the attack engine can continue flowing 500 gpm if the initial attack requires it. The supply and relay operators should not attempt calculations for flow rates under 500 gpm, even if the initial attack engine is flowing less than 500 gpm. This is because even if a single attack line is initially operating off of the attack engine, once a water supply is established, additional lines will more than likely be placed in service. In addition, the attack engine will also want to refill their booster tank to replenish the water used for the initial attack. All of this should result in a water flow at the scene in the 500 gpm range.

After the water supply is established, the operator should continue to keep track of the lines being placed in service at the fire scene. This knowledge will help the operator to effectively determine the supply needs for the incident. The operator should remember that supply operations are not a game of exact figures; the supply operator is not expected to get the flow rate down to the exact gallons per minute. For this reason, it is acceptable for the flow rates to be rounded in order to make the hydraulic calculations easier.

For example, if the fire scene has a 500 gpm deck gun, one 2½" line flowing 250 gpm, and one 1¾" exposure line flowing 150 gpm in service, the supply operator can estimate the flow rate to be around 900 gpm. Rather than do the supply calculations for 900 gpm, it is completely acceptable to round up to 1,000 gpm. Not only does this make the hydraulic calculations easy for the supply operator, but it also builds in a layer of redundancy at the fire scene.

To further simplify supply calculations for the supply operator, one recommendation is to increase the flow rate through the supply line by 500 gpm increments. As stated above, the starting flow rate for a supply operation should be 500 gpm. If it is determined that 500 gpm is not sufficient flow for the operation, the operator can increase the flow rate to 1,000 gpm. This not only keeps the hydraulic math simple, but it enables the fire scene to place twice as much water in service compared to the initial operation. If the fire scene does not need exactly 1,000 gpm at the scene, the attack engine will register a higher residual intake pressure. Depending on what the residual intake pressure at the attack pumper is, this may be advantageous for the attack engine. If the incoming pressure is too much, however, the attack engine should radio to the supply or relay pumpers to back down their discharge pressures.

If it is determined that more than 1,000 gpm flow is required at the fire scene, the supply and relay operators should be thinking about putting a parallel supply line operation in service. This will enable higher volumes of water to be delivered to the fire scene without ridiculously high pump discharge pressures and internal plumbing loss values. The exact process for deploying this type of operation will be discussed in chapter 9.

When we boil this information down to the pieces that are absolutely critical, the operator should recognize that there are really only two realistic flow rates that should be memorized: 500 gpm and 1,000 gpm. The operator should understand that supply operations should begin at a flow rate of 500 gpm for the reasons previously discussed. Due to losses in discharge plumbing and the realistic maximum distances for most supply lines, 1,000 gpm should be the next flow rate to memorize for single supply line operations. This makes determining the friction loss for these supply operations quite easy for the operator. The operator can simply remember the friction loss values listed in table 3–5 for the supply line being utilized.

Table 3–5. Simplified friction loss values for basic supply operations

Flow rate (gpm)	Single 3" (psi/100')	Dual 3" (psi/100')	Single 4" (psi/100')	Single 5" (psi/100')	Single 5" (DDZ values) (psi/100')
500	20	5	5	2	5
1,000	80	20	20	8	10

Note: Values are calculated using the coefficients for each hose, except for the last column, DDZ values. The DDZ values are calculated by the double drop zero method.

Tandem Pumping Operations

Scenarios may present themselves in which extremely high discharge pressures are required to deliver the appropriate volume of water. In many cases, the pressure required may be greater than a single engine is capable of producing on its own. A *tandem pumping operation* can be employed to achieve these high discharge pressures efficiently. Tandem pumping operations are essentially a closely located relay pumping operation between two pumpers.[8] The goal of this operation is to utilize two pumping apparatus to achieve the necessary pump discharge pressure for the operation.

Tandem pumping operations are typically the go-to tactic when dealing with very tall high-rise buildings. This tactic is also useful when delivering high volumes of water through aerial devices such as elevating platforms. As discussed previously, one of the major advantages of the centrifugal fire pump is its ability to make use of existing pressure. In a tandem pumping operation, this advantage is used to create high discharge pressures for the operation.

For example, consider a situation in which 300 psi is needed on the discharge of a pumper to supply crews operating on the top floor of a tall high-rise building. In order for a single pumper to produce 300 psi, the motor of the apparatus would have to operate at extremely high engine revolutions per minute. Even if the apparatus were capable of achieving this discharge pressure, the pump capacity of the apparatus would be severely limited. The solution to this problem is to implement a tandem pumping operation and utilize a second pumper to help build the necessary pressure.

During a tandem pumping operation, one pumper will be connected to the water source, and it will then have lines connected from its discharge outlets to the intakes on the second pumper. The second pumper will then run lines from their discharge outlets to the building's fire department connection (FDC). It is important to remember that these two pumpers will be located close to one another, typically within 100'. In the scenario outlined above, the first pumper will receive water from the source and discharge it to the second pumper at a discharge pressure of 150 psi. Depending on the friction loss in the hose, the residual intake pressure on the second engine should be about 150 psi. Because centrifugal fire pumps take advantage of incoming pressure, the second pumper only needs to add an additional 150 psi to the water in order for the pump discharge pressure to equal 300 psi (fig. 3–15). When tandem pumping operations are employed, we essentially create a two-stage pump in the pressure position utilizing two single-stage pumpers. This tactic allows the necessary pump pressures to be achieved utilizing two diesel motors compared to overworking a single apparatus.

Pump operators must be mindful of the fact that exceedingly high pump pressures are created during a tandem pumping operation. Operators must ensure that they do not exceed the operating pressure for the hose lines utilized for these operations. In high-rise scenarios, it is recommended that medium diameter hoses be utilized since they typically have higher operating pressures. When using tandem pumping operations to supply large volumes of water to elevating platforms, operators must ensure that they do not exceed the operating pressures for the large diameter supply lines being used. They should make every effort to avoid using discharges located on the driver's side of the apparatus. This will prevent a potential injury should a hose line burst.

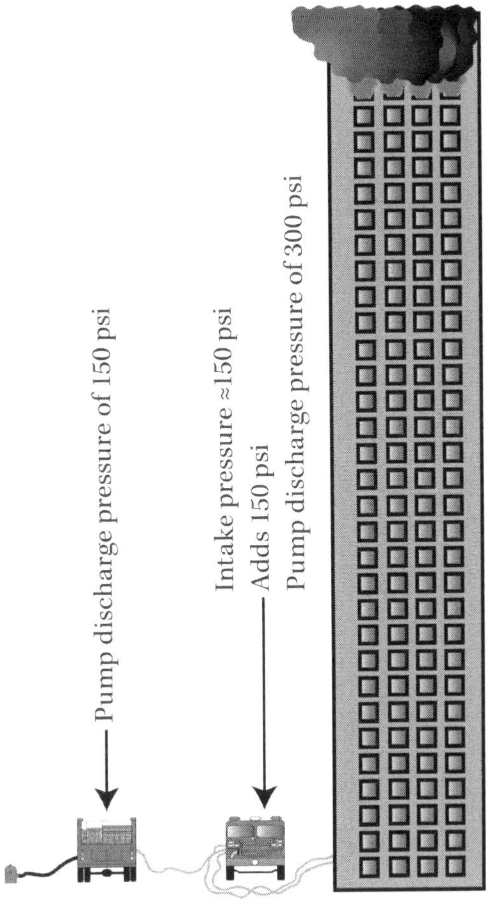

Figure 3–15. Tandem pumping

Dual Pumping/Intake-to-Intake Operations

Dual pumping operations have been a tactic utilized by fire departments since the early days of centrifugal fire pumps.[9] Some other names for this water supply tactic include *suction-to-suction*, *intake-to-intake*, or *low-side to low-side pump operations*. As each of these names implies, this tactic involves having two pumpers connected to one another via their intakes.[10] For operators who have never heard of this tactic, this seems like a completely incorrect pumping configuration. Although this water supply tactic is a highly beneficial option for supplying water to the fire scene, it is underutilized.

In order to understand the concept of intake-to-intake pump operations, the operator has to have a firm concept of residual pressure. Residual pressure refers to the pressure remaining in a system once water begins flowing. What this really means is that residual pressure is an indication of water in a system that is not being used. The higher the residual pressure, the more water is available within a pressurized system. A dual pumping operation essentially allows a second pumper to take advantage of the unused water entering the intake side of the first pumper. This is done by connecting the second pumper to the first pumper via an intake-to-intake connection.

For example, imagine Engine 1 is connected to a strong hydrant. Engine 1 is flowing 500 gpm through their deck gun, and their residual intake pressure is 80 psi. The residual intake pressure reading on Engine 1 indicates that there is more water available from the hydrant that Engine 1 is not using. If Engine 2 positions near Engine 1 and also needs to place their deck gun in service, a dual pumping operation can be implemented to supply Engine 2 with water (fig. 3–16). The second pumper should position within 100' of the first engine to minimize the friction loss in the supply hose connected between the apparatus.

After a supply line is connected between the intakes of both apparatus, Engine 1 can open the intake valve that is connected to the second engine. The residual pressure coming from the hydrant will force the unused water to flow through the intake plumbing of Engine 1 and directly into the intake of Engine 2. Once Engine 2 opens the discharge valve for their deck gun, the residual intake reading on both pumpers will drop. This is because both pumpers are now flowing water from the source. Engine 1 receives water first, takes what it needs, and then the unused water flows through its intake

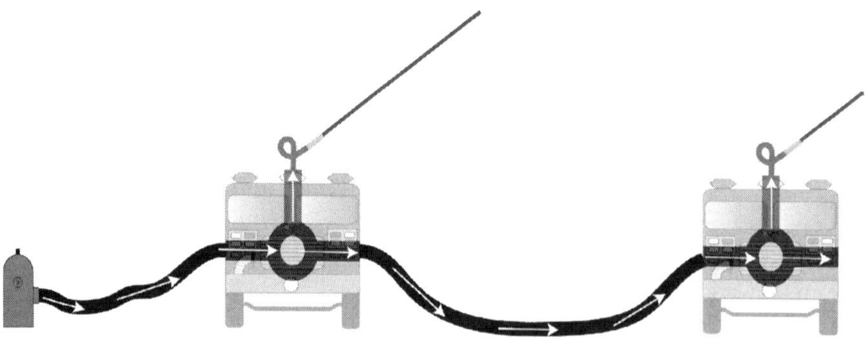

Figure 3–16. Dual pumping

manifold to feed Engine 2. The operator must remember in the scenario outlined above that once Engine 2 begins flowing water, the residual intake pressure will be lower on Engine 2's intake gauge compared to Engine 1's. This is due to the friction loss within the supply line and intake valves between the two pumpers.

When discussing the concept of intake-to-intake operations, operators often ask why this tactic should be employed instead of a traditional discharge-to-intake or relay configuration. The primary disadvantage of the traditional configuration is the likelihood of the second engine being unnecessarily overpumped. In order to prevent this, the first engine must attempt to gate back the discharge feeding the second engine. While this is possible, it adds a layer of complexity for the first engine operator. By performing a dual pumping operation instead, the potential of the second engine being overpressurized is eliminated. This is because the water feeding the second engine never passes through the spinning impeller, and thus no additional pressure is added. The major advantage to this configuration is that both engines have their intake valves and discharge valves fully opened, and both engines have full control of their throttles.[11] This arrangement will ultimately simplify the water supply operation.

In the dual pumping operation outlined above, the water flow is unidirectional from the source into the first pumper and then into the second pumper. If the first engine increases its flow rate without communicating with the second engine, there is a possibility that the second engine will be starved for water and will cavitate. This is because the second engine is essentially at the end of a dead-end water main. For this reason, the two operators must communicate their residual intake pressures with each other.[12] Remember, the residual intake pressure on the second engine will always read the lowest because this engine is the furthest downstream. Chapter 9 will discuss options for improving the efficiency of a dual pumping operation by implementing an aboveground looped system.

One instance where it is beneficial to use an intake-to-intake supply operation is when a high-flow fire hydrant is being utilized. These hydrants typically can supply more water than a single pumper has the ability to discharge to the fire scene. The limiting factor in these instances tends to either be the plumbing configuration of the apparatus or the horsepower of the apparatus motor (fig. 3–17). A dual pumping operation enables more of the capacity of a high-flow fire hydrant to be delivered to the fire scene. The operation will start with the first supply pumper connecting to the hydrant and supplying whatever it is capable of to the fire scene. If the first supply pumper reaches a point where they have run out of apparatus horsepower but still have a significant amount of residual intake pressure, the operator can conclude that the hydrant has more water available than a single pumper can supply. By implementing a dual pumping operation, the second supply pumper will have access to the unused water that the first supply pumper is not capable of pumping to the fire scene. In this scenario, the operators are essentially creating a two-stage pump in the volume position using two separate pumping apparatus.

A more common application of a dual pumping operation occurs when supplying additional pumpers at the actual fire scene. These scenarios typically unfold like any other supply operation. A supply line is laid between the water source and the attack engine, and a supply pumper is positioned at the source pumping to the attack engine. If a later-arriving pumper or quint aerial apparatus arrives and needs a water supply, a dual pumping operation can be established between the initial attack engine and the later-arriving rig. This will allow both rigs to have a water supply for the operation (figs. 3–18 and 3–19). The beauty of this operation is that it makes operations simple for all the units involved. The supply pumper

Figure 3–17. Two pumpers performing an intake-to-intake operation from one strong hydrant

Figure 3–18. The engine on the right is being supplied by a supply pumper operating at the hydrant. The later-arriving apparatus on the left is able to obtain a water supply via a dual pumping operation with the initial attack pumper.

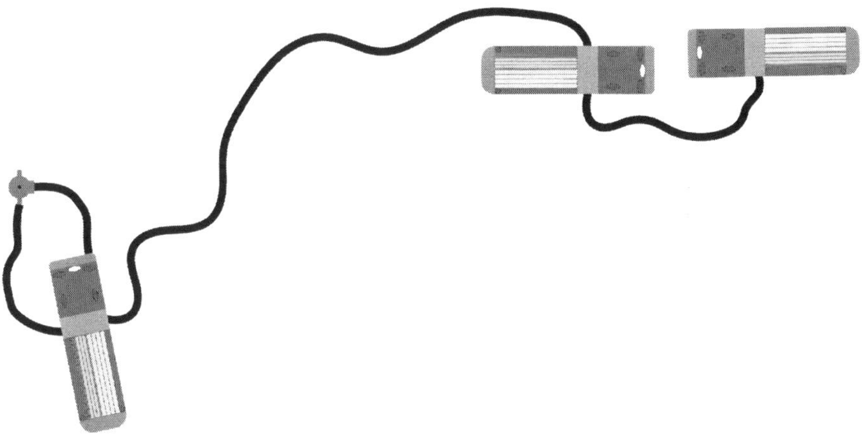

Figure 3–19. The engine at the hydrant is pumping to the initial attack engine, which is performing a dual pumping operation with a later-arriving engine company.

only needs to determine the total flow rate at the scene and perform the hydraulic calculations for that flow through the supply line. The water will naturally split between the two pumpers based on the flow demands of each apparatus, ultimately simplifying operations at the scene for the attack pumpers.

Booster backup operations can benefit enormously from dual pumping tactics. After the primary attack engine has connected the supply line to their intake and received water, a suction-to-suction operation should take place. The second engine should run a large diameter supply line from an intake on their apparatus to an unused large intake on the primary attack engine. Once this is accomplished, the primary attack engine can open their other intake valve and supply water to the second pumper. The second pumper can now refill their booster tank or place additional lines in service for the firefight. One of the major advantages of performing this operation at the scene is that the incident commander now has the ability to split the deployment of attack lines between two engines.

Intake valves are a critical piece of equipment that must be present on pumpers performing a suction-to-suction water supply operation. This will allow each operator to safely control the flow of water between the pumpers involved in the operation. Scenarios involving pumpers that have a blind-capped intake make it more difficult to perform dual pumping operations safely. This is due to the fact that residual pressure from the source causes the blind cap to be constantly under pressure. If the blind cap is removed while it is pressurized, a serious or fatal injury to the pump operator could result. In order to deal with this situation, operators formerly were taught to increase the flow rate on the initial apparatus until the residual intake pressure on that rig was decreased to somewhere between 0 psi and 5 psi.[13] The operator could then pop the blind cap off the intake and connect hose between pumpers. *Pump operators must understand that this practice is not recommended and is extremely dangerous.* The safest and most efficient way to conduct a dual pumping operation is to utilize intakes equipped with intake valves on both pumpers involved in the operation.

The operator should remember that suction-to-suction operations can be used in both urban and rural environments. Urban applications typically involve the supply engine pumping water from a hydrant to a fire scene where two or more pumpers are performing a dual pumping operation. In the rural environment, the supply pumper will typically be drafting from a dump tank and supplying water to the scene in the same fashion. Regardless of the type of source, dual pumping operations require water to be pressurized in order for the process to take place. Because the supply engines are pumping water to the scene in both instances, suction-to-suction operations are possible between the engines at the fire scene.

Study Questions

1. Which of the following is the most realistic option for increasing the volume of water available to the supply pumper?
   a. Increase the velocity of the water
   b. Increase the area feeding the intake side of the pump
   c. Decrease the velocity of the water
   d. Decrease the area feeding the intake side of the pump

2. Which of the following is true regarding a booster backup operation?
   a. It is only useful in the rural environment.
   b. It requires the initial attack engine to perform a forward lay from the water source.
   c. It is intended to extend the operational time of the initial attack engine while a sustainable water supply is established.
   d. It is intended to prevent the need to utilize an external water supply.

3. Flowing 1,000 gpm, the friction loss in ____________ 2½" supply lines is the same as a that of a single 5" LDH.
   a. Two
   b. Three
   c. Four
   d. Five

4. According to *NFPA 1962*, hose marked SUPPLY HOSE should not be operated at pressures that exceed:
   a. 250 psi
   b. 200 psi
   c. 185 psi
   d. 150 psi

5. Which of the following is the minimum fire flow rate that operators of supply pumpers should perform their initial pump calculations for?
   a. 150 gpm
   b. 300 gpm
   c. 500 gpm
   d. 1,000 gpm

6. Generally speaking, most apparatus deck guns are capable of flowing between:

    a. 500–1,000 gpm
    b. 1,000–1,500 gpm
    c. 1,000–2,000 gpm
    d. 1,250–1,750 gpm

7. Tandem pumping operations are intended to do which of the following?

    a. Connect two pumpers intake-to-intake in order to share a strong water source
    b. Connect two pumpers discharge-to-discharge in order to share a strong water source
    c. Connect two pumpers intake-to-intake in order to build high pressures
    d. Connect two pumpers discharge-to-intake in order to build high pressures

8. Tandem pumping operations are similar to a two-stage pump operating in which position?

    a. Volume
    b. Pressure
    c. Rpm
    d. Parallel

9. Which of the following pieces of equipment will be required to safely perform a dual pumping operation?

    a. Hose clamp
    b. Dump tank
    c. Intake valve
    d. Four-way hydrant valve

10. Which of the following is true regarding dual pumping operations?

    a. They are the same thing as tandem pumping operations.
    b. They cannot be utilized in the rural environment.
    c. Residual pressure on the first pumper is used to supply a second pumper via a discharge-to-intake connection.
    d. Residual pressure on the first pumper is used to supply a second pumper via an intake-to-intake connection.

4

DRAFTING OPERATIONS

The ability to conduct successful drafting operations is one of the most fundamental skills required of a proficient pump operator, but many operators avoid drafting operations at all costs. There are varied reasons why pump operators and fire departments avoid drafting operations. Some departments use the excuse of having hydrants installed in close proximity, perhaps every 200' or so, as a reason for not being proficient with drafting operations. Other departments cite the need for personnel and equipment as the reason for their avoidance of drafting operations. Yet other pump operators evade drafting operations simply due to a lack of confidence in their ability to achieve a reliable prime.

Engineers must understand that the most basic function of the fire department pumping apparatus is its ability to get water from a static water source. While it is true that many urban fire departments do not routinely find themselves needing to employ this water supply tactic, it certainly does not eliminate the possibility of these departments having to draft within their city. Several instances have occurred across the United States where urban fire departments have had to implement drafting operations because of a failure in their municipal water system. It is also a real possibility for urban fire departments to be forced to implement rural water supply tactics in conjunction with municipal water sources in order to achieve the needed fire flow for larger incidents.

In this chapter, we will examine the basic principles related to drafting operations. This chapter will also discuss alternative drafting methods that do not require the use of an onboard mechanical primer and explore when these methods should be employed. Options for maximizing the flow capabilities of pumpers operating from static sources will also be discussed. By the conclusion of this chapter, the pump operator should have a thorough understanding of the major concepts related to drafting operations and should recognize that the art of drafting is not as challenging as some believe.

Drafting Physics

Although the pump operator does not need to be a physicist in order to successfully operate from a draft, it is important for the operator to have a thorough understanding of the science related to the practice. Some firefighters think of a drafting operation as "sucking water into the pump." This common misconception is not a true description of what actually occurs when operating from a static water source. Drafting occurs when the pressure at the eye of the impeller is lowered below the atmospheric pressure outside of the pump. This is known as a *vacuum within the pump* and is typically achieved using the onboard mechanical primer. At sea level, the atmospheric pressure is 14.7 psi. When a vacuum is created on the intake side of the pump, the water from the static source is pushed into the pump by the atmospheric pressure acting on it.[1] The pump is considered to be primed once the water has been pushed up the hard sleeve and into the eye of the impeller by the atmospheric pressure.

The most common cause of failure during a drafting operation is an air leak of some sort. This usually occurs either because the hard sleeve connections are not tight enough or a drain is accidentally left open. Prior to connecting sections of hard sleeve, it is critically important that firefighters look inside the female swivel connection and ensure a gasket is present. If threaded sections of hard suction hose are connected without a gasket present, the seal between the two sections will never be tight enough to allow a prime to be achieved. After confirming the presence of a gasket, the operator must ensure that the connection is tightened sufficiently. This requires the operator to take a mallet to the connection to ensure that it is airtight. Failure to perform these two critical steps will prevent the pump from achieving a prime. In addition, the time wasted chasing air leaks on the rural fireground will result in delayed application of water on the fire scene.

Some fire departments elect to utilize hard suction hose equipped with Storz couplings instead of traditional sexed couplings. When using hard sleeve with Storz fittings, it is critically important that the pump operator ensure that the proper Storz gasket is present for the drafting operation (figs. 4–1 and 4–2). Pressure gaskets only seal when the hose and fittings are subjected to positive pressures, whereas Storz drafting gaskets create an airtight seal during drafting operations. If two Storz-equipped hard sleeves with pressure gaskets are used during a drafting operation, a prime will never be achieved because the gaskets are not capable of sealing when subjected to a vacuum. When using hard sleeves with Storz fittings, it is recommended that one coupling contain a drafting gasket while the other contain a pressure gasket. The rib of the drafting gasket will create an airtight seal against the other pressure gasket to ensure a prime can be achieved. If no Storz drafting gaskets are present or available, as a last resort, the operator can spread petroleum jelly or some other vacuum-distilled product around the surface where the two fittings meet when they are connected. The operator should then wrap the entire connection in duct tape to help create an airtight seal (fig. 4–3).

When discussing drafting operations, the operator must understand the difference between *gauge pressure (psig)* and *absolute pressure (psia)*. Gauge pressure refers to the reading on a pressure gauge without taking into

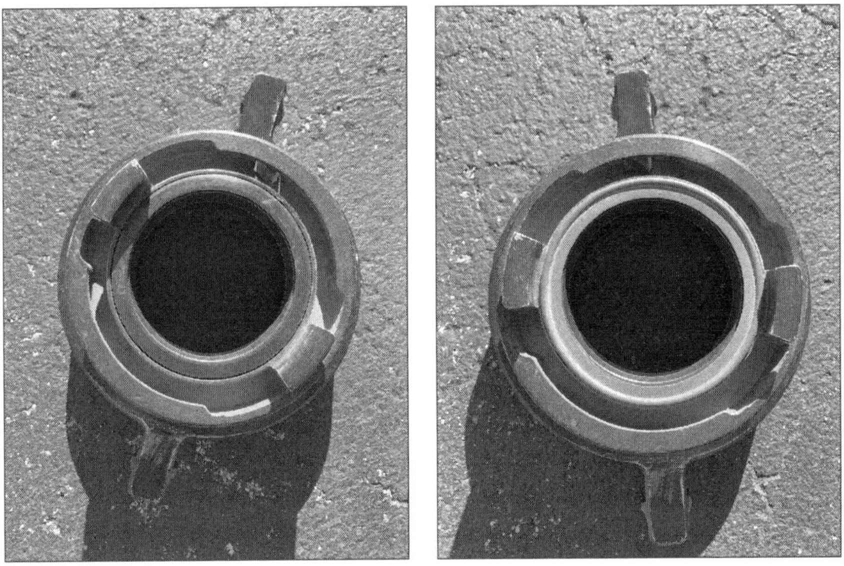

Figure 4–1 and 4–2. A Storz fitting equipped with a pressure gasket (left). A Storz fitting equipped with a drafting gasket (right). Note the rib on the outer edge of the gasket. One drafting gasket must be present in the Storz coupling when attempting to draft through these fittings.

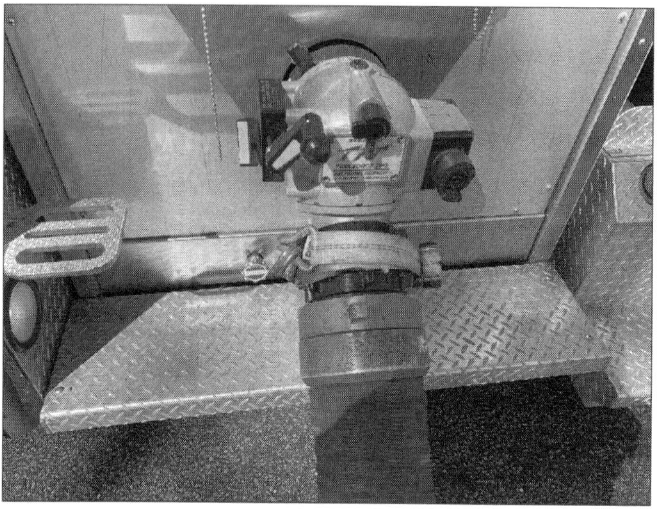

Figure 4–3. If the proper Storz gaskets are not available for a drafting operation, the operator can spread vacuum-distilled jelly around the mating surface of the couplings and wrap it with duct tape. The ratchet strap was added in this photo as an extra measure to ensure the seal remained airtight.

account atmospheric pressure. Absolute pressure is the actual reading on a pressure gauge after atmospheric pressure has been factored in. Pump operators must remember that the gauges on their pumping apparatus read in terms of pounds per square inch gauge. Thus when the pressure gauge reads 0 psi, the absolute pressure is actually 14.7 psi at sea level. To reach an absolute pressure of 0 psi would require a perfect vacuum to be created. Centrifugal fire pumps are not capable of producing a perfect vacuum because of the tolerances that must be maintained in order for the impeller shaft to spin in the pump casing. If these tolerances are too loose, a prime will not be achieved because too much air will be capable of entering the pump. If the clearance between the spinning impeller shaft and the pump casing is too tight, however, the impeller shaft will overheat while it spins.[2]

Operators should remember that as the elevation above sea level increases, atmospheric pressure decreases. This is because there is less air above the water and therefore less pressure available to push water into the fire pump at higher elevations. Atmospheric pressure decreases approximately 0.5 psi for every 1,000' rise in elevation.[3] Generally speaking, the impacts of operating at higher elevations do not cause significant issues until the elevation above sea level is greater than 2,000'. Conversely, areas that are below sea level will experience higher atmospheric pressures. In addition, changes in weather patterns will also have an effect on the atmospheric pressure for a given location.

Lift

While operating from a draft, atmospheric pressure is utilized to overcome two variables: *lift* and friction loss within the hard sleeve/strainer. Lift is defined as the difference in elevation between the centerline of the pump and the surface of the water. As lift increases, more atmospheric pressure is utilized to overcome gravity in order to push water into the pump. This means that there is less atmospheric pressure available to overcome friction loss in the hard sleeve. Therefore, as the lift is increased, the flow rate that is achievable from a draft is dramatically decreased. Operators must remember that there is only a certain amount of atmospheric pressure available to overcome these two variables. If a higher flow rate is desired, there must be less lift. If there is a larger amount of lift to overcome, the operator should expect a lower flow rate (fig. 4–4).

The pump operator must be familiar with the maximum height at which the pumping apparatus can be located above the water source for a drafting operation. The elevation above sea level will have a direct impact on how high the pumper can be located above the water source due to the reduced amount of atmospheric pressure at higher altitudes. If a

Figure 4–4. This engine is drafting through a twin tube setup from a lift of more than 20'. The lift present reduced the achievable flow rate to 900 gpm.

centrifugal pump were capable of creating a perfect vacuum, water could be lifted to a theoretical height of 33.9' at sea level (fig. 4–5). This is known in the fire service as *theoretical lift*. Since we know that centrifugal fire pumps are unable to achieve a perfect vacuum, theoretical lift is not something that the pump operator should ever expect to overcome.

The pump operator should be more concerned with the concept of *maximum lift*. In the fire service, maximum lift is defined as the highest height that water can be lifted by a centrifugal fire pump under real-world conditions. This again will vary based on weather conditions and the elevation above sea level. For firefighting purposes, maximum lift is considered to be roughly 25' at sea level. This means that a pumper can be a maximum of 25' above the water source and be expected to lift water to the pump.

It is critically important for the pump operator to understand that when the apparatus is 25' above the water source, almost all of the atmospheric pressure available is being utilized to overcome gravity. This means that little, if any, atmospheric pressure is available to overcome the friction loss in the hard sleeve/strainer, and thus the achievable flow rate will be less than ideal. For these reasons, the pump operator should avoid using water sources that are 25' or more below the pumping apparatus.

Ideally, the pumper should be positioned in a fashion that minimizes lift. This allows more of the atmospheric pressure to be used to overcome the friction loss in the hard sleeve and not lost overcoming gravity. Since it is not always realistic to completely eliminate lift from a scenario, the pump

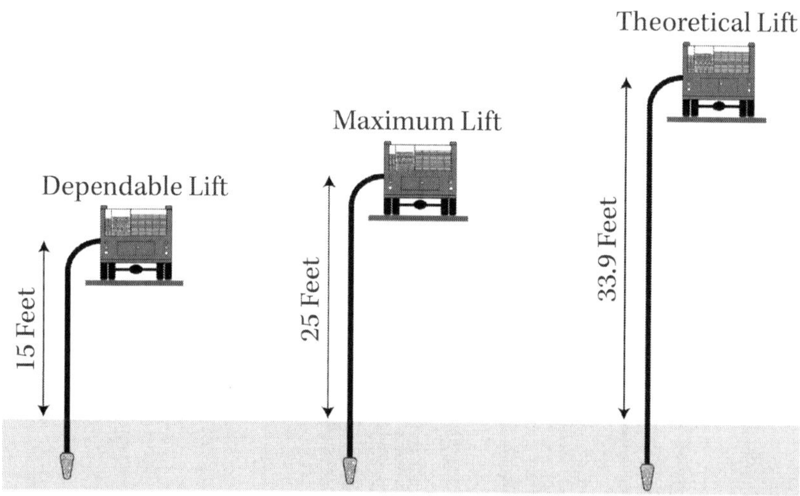

Figure 4–5. Dependable, maximum, and theoretical lift

operator must know the maximum height at which the apparatus can be located above the water and still achieve a reliable flow rate. This height is known as *dependable lift*, and at sea level, it is generally considered to be around 15'. When the pumper is 15' above the water source, there is a sufficient amount of atmospheric pressure available to overcome the lift and also to overcome the friction loss in the hard sleeve/strainer, allowing a reliable flow rate into the pump. While identifying and preplanning rural water sources, fire departments should ensure that there is not more than 15' of lift between the location where the pumper can be safely positioned and the surface of the water. This ensures that a sufficient volume of water can be delivered from these sources for rural water supply operations.

The operator must understand that there is a difference between a reliable flow rate and the rated capacity of the pump. According to *NFPA 1900*, centrifugal fire pumps must be capable of flowing their rated pump capacities under certain lift requirements. The maximum lift that is allowed is dependent on the rated capacity of the pump.[4] Pumps rated up to 1,500 gpm must be capable of achieving this flow rate with no more than 10' of lift. For 1,750 gpm pumps, the maximum allowable lift is 8', and for 2,000 gpm and larger pumps, the maximum allowable lift is 6'. It becomes evident that as the rated capacity of a pump increases, the maximum allowable lift decreases in order for the rated pump capacity to be achieved. This is because higher flow rates require more of the available atmospheric pressure to be utilized to overcome the friction loss in the hard sleeve. Any substantial lift will prevent these higher flow rates from being achieved.

When using static water sources for fill sites in the rural environment, the goal should always be to achieve at least a 1,000 gpm flow rate from the source. This enables mobile water supply apparatus to be filled at the *NFPA 1900* fill rate of 1,000 gpm.[5] While it is unlikely that a pumping apparatus will be capable of achieving its maximum rated capacity from a source that is 15' below the pumper, a 1,000 gpm flow rate is achievable when the proper drafting techniques are implemented (fig. 4–6).

Horizontal reach

The lift, or vertical distance between the surface of the water and the centerline of the pump, has a significant impact on the available flow rate during a drafting operation. One common drafting misconception held by pump operators is that they can only draft through a maximum of 20' of hard sleeve. *NFPA 1900* requires pumpers to carry a minimum of 20' of hard

Figure 4–6. Each of these pumpers is flowing 1,000 gpm while overcoming approximately 15' of lift.

sleeve or 15' of supply hose.[6] Because of this, many engine companies in the United States only carry two 10' sections of hard sleeve. Therefore, some operators believe this is the maximum amount of hard sleeve that they are capable of drafting through.

In addition to the maximum lift requirements set forth by *NFPA 1900* and *NFPA 1910*, the standard also specifies the length for hard sleeve attached to each intake during pump certification testing. These standards require pumpers to flow their capacity through 20' of hard sleeve for each intake that is permitted in order for the pump to reach its capacity.[7] This means that no single intake will ever have more than 20' of hard sleeve connected to it during a pump certification test. Operators must remember that pump certification tests are intended to ensure that a pumper can deliver its rated pump capacity under ideal conditions. In the real world, operators will often find themselves having to reach water that is more than 20' away from where the pumper can safely position (fig. 4–7).

We know that the maximum lift at sea level is 25' vertically. At this height, all the atmospheric pressure is being used just to overcome gravity. This understanding causes many pump operators to confuse the principles of lift with the principles of *horizontal reach*. When operating horizontally with minimal lift, the atmospheric pressure available is used to overcome the friction loss in the hard sleeve for a given flow rate. For these reasons, it is possible to connect more than 20' of hard sleeve together in a *long reach*

Figure 4–7. Situations often occur where the water source is more than 20' from the pumper.

operation to reach a distant water source and still obtain a respectable flow rate.

One of the limiting factors of a long reach operation is the friction loss in the hard sleeve for a given flow rate. Pump operators should understand that they can connect double or triple the amount of hard sleeve together, however, and achieve a higher flow rate than when operating 25' above the surface of the water. When performing a long reach operation, the operator must be mindful of the lift over the entire horizontal run of hard sleeve. In other words, it is possible to draft a reliable flow rate through 80' of hard sleeve as long as the total lift remains minimal (fig. 4–8). Atmospheric pressure is used to overcome both lift and friction loss in the hard sleeve. Of these two variables, lift has a much bigger impact on the operational capabilities of the apparatus.

Cavitation

As mentioned previously, it is important to prevent cavitation from occurring during pumping operations. Cavitation can occur when a pump operator is attempting to discharge more water than the pump is receiving. Operators often recognize cavitation by the sound the pump makes when it occurs, which sounds like the pump is discharging gravel rather than water. Cavitation has the potential to cause serious damage to the pump, and the pump operator must be able to recognize its warning signs and be familiar with the actions that should be taken to avoid its occurrence.

Figure 4–8. This pumper is flowing 1,000 gpm through 80' of 6" hard sleeve. This was possible because there was minimal lift.

In order to fully understand what is occurring when a pump is cavitating, we must look at this phenomenon from a scientific perspective. Cavitation occurs when the pressure on the intake side of the pump drops below the vapor pressure of the water being drafted.[8] When this occurs, the water entering the pump transitions from a liquid into gaseous vapor bubbles. As these vapor bubbles pass through the pump, they are subjected to the high pressures within the pump and implode, which causes them to transition back into a liquid state of matter. This rapid change between states of matter is extremely destructive to the pump, and the operator should avoid reaching this point.

Cavitation most commonly occurs while operating from a draft. Potential causes of cavitation include the following:[9]

1. Too much lift between the water source and the pump
2. Too much water being discharged from the pump
3. A strainer that is either partially or completely clogged or blocked
4. Collapse of the hard sleeve (fig. 4–9)
5. Use of suction hose that is too small for the desired flow rate
6. Drafting from water sources that contain warm or hot water

As stated in the previous section, lift should be minimized as much as possible during drafting operations. This is because most of the atmospheric

Figure 4–9. Heat from the exhaust pipe of this pumper weakened this section of hard sleeve sufficiently to cause it to collapse.

pressure acting on the water is used to overcome gravity in these scenarios, which naturally results in lower flow rates prior to reaching cavitation. When dealing with sources that have a lower lift, cavitation is still possible. In order to reach that point, however, higher volumes of water must be flowed. In order to achieve the higher flow rates, the pump must create more of a vacuum to overcome the friction loss in the hard sleeve. This is why higher flow rates can be expected at lower lifts prior to the point of cavitation.

Many of the water sources used in the rural environment contain debris and vegetation that are capable of clogging the strainer at the end of the hard sleeve (fig. 4–10). When the strainer becomes clogged with debris, the area of the strainer exposed to the surrounding water is diminished. This will require a larger vacuum to be created on the intake side of the pump in order to maintain the flow rate. The likelihood of the pump reaching cavitation in these scenarios will depend on how badly the strainer is clogged.

The pump operator must also watch to ensure none of the hard suction hoses utilized for the drafting operation begins to collapse. There are many variables that can cause a hard sleeve to collapse during a drafting operation. The most common cause occurs when the hard sleeve is positioned too close to the exhaust pipe of the apparatus. Heat exiting the exhaust pipe is sufficiently hot during a pumping operation to warm up the hard sleeve and cause it to weaken and collapse, which can result in cavitation of the

Figure 4–10. Sources with a lot of debris could cause the strainer to clog.

pump. Another cause of collapsing hard sleeve occurs when it is draped over sharp 90° corners, such as edges of bridges or dump tanks. This issue is amplified when lightweight hard sleeve is used in these situations on sweltering summer days. The weight of the water in the hard sleeve coupled with the flexibility of the hose itself can cause the hard sleeve to collapse, occluding water flow to the pump (fig. 4–11). This will rapidly cause the pump to cavitate.

While operating from a draft, pump operators should pay attention to the master intake gauge because it can inform them of the possibility of cavitation. If the master intake gauge starts reading a higher vacuum reading without an increase in the flow rate, either a blockage has begun to form at the strainer or the hard sleeve itself has begun to collapse. Both of these scenarios will cause a higher vacuum reading on the compound gauge because the pump must work harder to flow the same volume of water through the occlusion. The pump operator should swiftly determine which of these issues is occurring and take action to remedy the problem.

As stated in previous chapters, the area of the opening has the biggest impact on the volume available during any supply operation; the larger the hose, the less friction loss at a given flow rate. For drafting operations, it is advisable to utilize 6" hard sleeve instead of 5" sleeve because there is less friction loss at a given flow rate in the 6" hard sleeve. This means that 6" hard sleeve will enable the pump operator to achieve higher flow rates before reaching cavitation as opposed to 5" hard sleeve.

Figure 4–11. The weight of the water in the hard sleeve and the angle at which it is bent have the potential to cause this lightweight hard sleeve to collapse.

The temperature of the water being drafted has a significant impact on the likelihood of a pump reaching cavitation. As the temperature of the water increases, so does its vapor pressure, which means that the pump will reach cavitation sooner. For this reason, the pump operator must remember that cavitation will occur faster when drafting warm water compared to cold water. This is the primary reason the water temperature should remain between 35°F and 90°F during an annual pump test.[10] If the water temperature rises above 90°F, the pump will likely cavitate before it is able to reach its rated pump capacity, which will affect the results of the test.

When the water temperature gets too high during an annual pump test, the drafting pit or dump tank utilized can be drained and refilled with cooler water. On a fire scene, however, this is usually not an option. A pump operator who is faced with drafting from a source that contains warm water should attempt to minimize the lift between the pumper and the source as much as possible. This will enable the pump to flow more of its capacity before reaching cavitation. The operator should also remember that stagnant water sources within the dependable lift range may yield a lower flow rate on a sweltering summer day (fig. 4–12). This is due to the fact that often the water in these sources is constantly exposed to sunlight, causing the water temperature to rise dramatically. Ideally, flowing sources of water should be utilized for drafting operations during the summer months instead of stagnant sources since the temperature of the water will be lower.

Figure 4–12. Stagnant water sources may yield lower flow rates on hot summer days due to the temperature of the water.

All of these variables influence the likelihood of the pump reaching cavitation, and pump operators must be cognizant of how close they are to reaching this point. A major precursor to cavitation is known as "running away from water," as discussed previously. This occurs when the pump operator increases the revolutions per minute of the engine with no corresponding increase in discharge pressure. Once this point is reached, further increases in the revolutions per minute of the engine will actually result in a drop in discharge pressure. If the operator does not recognize this and continues to increase the throttle, the pump will reach cavitation.

Some obvious steps for preventing the pump from cavitating include lowering the lift, using hard sleeves with a larger internal diameter, and reducing the number of sections used for the operation. While these are valid options, certain scenarios may not allow these steps to be taken. If these variables cannot be changed or improved, the pump operator should take the following steps in order to prevent cavitation from occurring:

1. Reduce the flow rate from the apparatus.
2. Reduce the revolutions per minute of the engine.
3. Clear any blockages around the strainer.
4. Place an additional intake in the water to increase the total area feeding the pump.
5. Immediately replace collapsed or damaged sections of hard sleeve and make adjustments to prevent the new sections from collapsing.

Understanding the compound gauge

The master intake gauge is considered a compound gauge because it is capable of reading both positive and negative pressures, as discussed in chapter 1. The positive pressure side of the scale obviously reads in psi, while the negative, or vacuum, portion of the scale is represented in inches of mercury (inHg).[11] Although the vacuum portion of the compound gauge is smaller than the pressure side of the gauge, it can provide the pump operator with a tremendous amount of information during a drafting operation. This information will enable the pump operator to efficiently maximize the amount of water obtainable from a static water source.

During a drafting operation, the compound gauge will read in the vacuum portion of the gauge. The inches of mercury reading on the gauge will inform the operator concerning how much of a vacuum the pump is creating and therefore the amount of work done on the intake side of the pump. When evaluating the reading on the compound gauge, the pump operator must understand the difference between the *static intake reading* and the *dynamic intake reading* on the gauge. The static intake reading refers to the vacuum reading on the compound gauge after the pump has been primed, but before water is being discharged. This represents the vacuum required to overcome the lift between the water source and the pump.[12] Using the lift equation below, the pump operator can determine how high the water has been lifted based on the static intake reading:

$$L = 1.13 \times P_{\text{inHg}}$$

where

L = Lift in feet

1.13 = Feet of head equal to one inch of mercury (inHg)

P_{inHg} = Pressure, read as inches of mercury on the compound gauge

Example 4–1. You are operating the supply pumper from a draft. After priming the pump, you notice that the compound gauge is showing a static intake reading of 12 inHg. How much lift did the pump overcome?

$L = 1.13 \times P_{\text{inHg}}$

$L = 1.13 \times 12$

$\mathbf{L = 13.56'}$

In example 4–1, the static intake reading tells the operator that the pump needed to develop a vacuum of 12 inHg in order to lift the water 13.56'. It is important to remember that the gauges on fire service apparatus read pressure in pounds per square inch gauge and do not take into account atmospheric pressure. Assuming we are at sea level, in order to determine the absolute pounds per square inch of pressure for this scenario, we would have to first convert the inches of mercury reading on the compound gauge into pounds per square inch. Then we would subtract it from 14.7 psi. The following equation can be utilized to convert pressure in inches of mercury into pressure expressed as pounds per square inch (P_{psi}):

$$P_{psi} = 0.489 \times P_{inHg}$$

where

P_{psi} = Pressure, pounds per square inch (psi)
0.489 = Amount of pressure in psi equal to pressure in 1 inHg
P_{inHg} = Pressure, inHg

Example 4–2. Determine the absolute pressure per square inch from example 4–1 when the compound gauge is reading a static intake pressure of 12 inHg.

$P_{psi} = 0.489 \times P_{inHg}$
$P_{psi} = 0.489 \times 12$
$P_{psi} = 5.87$
$P_{psia,\ draft}$ = 14.7 – static intake reading
$P_{psia,\ draft} = 14.7 - 5.87$
$\boldsymbol{P_{psia,\ draft} = 8.83}$

Example 4–2 shows that the pressure on the intake side of the pump is actually 8.83 psi when the compound gauge reads 12 inHg. Since this is less than atmospheric pressure, we know that the pump has created a vacuum. While the operator would likely never make this calculation from the compound gauge while on the fireground, this example helps to show the pump operator what the inches of mercury values on the compound gauge actually represent.

Although the static intake reading provides the operator with valuable information concerning lift, it does not paint the total picture. Pump operators should be more concerned with the dynamic intake reading on the compound gauge. The dynamic intake reading is the reading on the compound gauge after a prime is achieved and the pumper is discharging water.[13]

This reading takes into account all the loss factors that the pump is overcoming in order to discharge at the current flow rate. These losses include the lift present, the friction loss in the hard sleeve and intake strainer, and the velocity head that is required to accelerate the water from a standstill. By paying attention to the dynamic intake reading, the pump operator will be able to determine in real time whether or not the current drafting configuration is maxed out.

Example 4–3. You are operating the supply pumper from example 4–1. After achieving your prime, your static intake reading is 12 inHg, and you determine you have overcome 13.56' of lift. You begin discharging water from the pumper and notice that your dynamic intake reading is 22 inHg. Using the lift equation, determine the lift for the dynamic intake reading.

$$L = 1.13 \times P_{\text{inHg}}$$
$$L = 1.13 \times 22$$
$$\mathbf{L = 24.86'}$$

Example 4–3 shows that 22 inHg is the required amount of vacuum to overcome a lift of approximately 25'. The actual distance between the pumper and the surface of the water in this example is only 13.56', however. This means that in order to overcome the friction loss in the hard sleeve for a given flow rate and 13.56' of lift, the pump is performing the equivalent amount of work required to lift water 25' (fig. 4–13). Since we know that maximum lift is 25', we can conclude that the pumper in this example has maximized its current drafting configuration. The operator should remember that a static intake reading of 22 inHg means that the pumper is exactly 25' above the water source. But when the dynamic intake reading shows 22 inHg, it means that all of the combined loss factors are causing the pump to do the same amount of work required to overcome 25' of vertical lift. If the pump operator attempts to flow more water once the dynamic intake reading reaches 22 inHg, the pump will likely cavitate.

It is important to note that some pumping apparatus may be capable of achieving a vacuum reading greater than 22 inHg, especially when they are new. Even so, this capability will likely diminish due to normal wear and tear as the pump ages. Pumpers are expected to be capable of achieving 22 inHg for the duration of their service life based on NFPA standards and the acceptable tolerances previously discussed. A vacuum test is performed as part of an annual *NFPA 1910* pump test to ensure that the pumper is

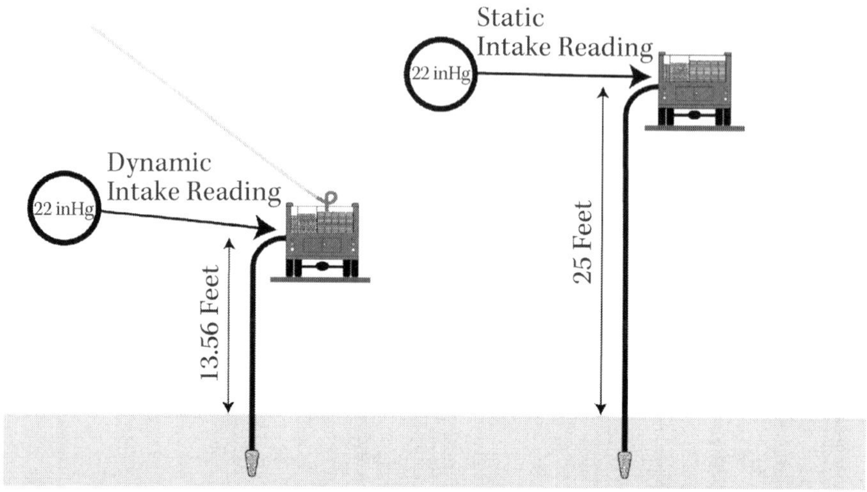

Figure 4–13. The dynamic intake reading of 22 inHg on the left pumper represents the vacuum required to overcome 13.56' of lift and flow water. Since the pumper on the right is 25' above the water, the same 22 inHg will have to be created just to lift water to the pump. Therefore, both pumpers are doing the same amount of work on the intake side of the pump, but only one of them can produce a reliable stream.

capable of reaching 22 inHg.[14] Operators should remember that while newer apparatus may be capable of achieving a vacuum greater than 22 inHg, no pumper is capable of achieving a perfect vacuum, which is the equivalent of 30 inHg. Because 22 inHg is the maximum vacuum reading that pumpers are required to achieve during an annual pump test, we will use this value to signify when pumpers have reached their maximum capability in the field as well.

Depending on the graduation of the scale on the compound gauge, it may be difficult for the operator to decipher a reading of 22 inHg while drafting. Many compound gauges found on modern fire apparatus show the vacuum portion of the gauge in increments of 10 inHg. On these compound gauges, the pump operator should consider the drafting configuration to be maxed out when the dynamic intake reading shows 20 inHg. Departments that routinely draft should consider outfitting their pumpers with a 50/50 style compound gauge (fig. 4–14). The positive pressure side of these gauges are one-half of the gauge's scale, while the vacuum portion of the gauge is the other half of the scale. These gauges enable the operator to have an extremely accurate idea of the dynamic intake reading during a drafting operation.

Figure 4–14. This 50/50 style gauge allows the operator to have an accurate idea of what the vacuum reading is during a drafting operation.
(*Courtesy*: Mike Guzy)

During a drafting operation, the pump operator should be constantly watching the dynamic intake reading to determine the efficiency of the operation. As mentioned, when this reading approaches 22 inHg, the drafting configuration is approaching its maximum capacity. This does not necessarily mean that more water is not available from the source. It simply means that the current drafting configuration is maxed out. Assuming that there is not a blockage at the strainer, this typically occurs because there is too much lift and friction loss in the hard sleeve present for the current flow rate. In order to either maintain or exceed the current flow rate, a twin or triple tube drafting operation should be performed.

Conventional Drafting

Modern fire apparatus typically utilize either a *mechanical primer* or an *air primer* to create a vacuum for drafting operations due to the fact that centrifugal fire pumps are not self-priming. Mechanical primers are typically either *rotary gear pumps* or *rotary vane pumps* that are powered by a small electric motor. Air primers utilize air supplied from the chassis air brake system. Regardless of the type of primer utilized, the basic operating principles are the same. The use of an onboard primer will typically achieve a prime within 15 to 20 seconds. Activating the primer for periods longer than this typically indicates that there is an air leak somewhere in the system. In addition, running a mechanical prime for too long poses a risk of burning up the electric motor.

Once the necessary connections are made, the intake valve should be opened, and the throttle increased before activating the primer. A prime has been achieved once the master discharge gauge begins reading pressure and the primer pump sounds as if it is bogging down. Once this point has been reached, the operator should slowly open a discharge and gradually increase the throttle to around 100 psi to ensure a prime has been achieved. When a prime is established, the operator must remember to continuously flow water. Failure to do so could result in a loss of prime and an interruption in water delivery.

Mechanical primers are highly efficient at creating the vacuum necessary for priming a centrifugal fire pump. If the pump is completely empty, the primer will be the only option for getting water from the outside static source into the fire pump. On many apparatus, however, primer pumps are not sufficiently employed and often fail to operate at rural fire scenes. These instances can prove to be extremely problematic on the rural fireground. The sections that follow highlight alternative drafting methods that do not require the use of the onboard primer for situations in which these devices fail to operate.

Alternative Drafting Methods

When onboard priming devices fail to operate, it can be highly stressful for the unprepared pump operator. These situations will require the pump operator to perform some type of alternative drafting technique. Alternative drafting techniques also become extremely beneficial when the operator is attempting to establish a multiple intake drafting operation without losing the initial prime.

As stated in chapter 1, centrifugal fire pumps are not considered to be self-priming and thus must be primed when they are completely void of water. In these situations, a positive displacement pump is the only option for creating an area of negative pressure on the intake side of the pump in order for the pump to be primed. If the pump casing and booster tank are full of water, however, several options exist for achieving a prime from an outside source without activating the onboard priming device.

Most fire departments run what are considered to be *wet pumps* because the pump casing is full of water prior to the incident and is therefore primed. When given a drafting assignment, a wet pump enables the pump operator to utilize alternative drafting methods to establish a prime from an outside static source. In order for any of these techniques to work, the following components must be present and in operation:[15]

1. The pump casing and booster tank must contain water.
2. The intake being primed must be equipped with an intake valve.
3. Water must be in motion at a relatively high velocity.

The pressurized prime

The *pressurized prime* is an alternative drafting method that utilizes the built-in jet siphon attachment found on many low-level strainers (fig 4–15). Prior to placing the strainer into the water source, the operator connects a 1¾" hose to the jet siphon attachment on the strainer. This process requires a portion of the booster tank water to be discharged through this 1¾" line and the jet siphon attachment on the strainer. Once the water is sent through the jet siphon at the appropriate pressure, a *venturi*, or area of negative pressure, is created that draws water from the source up the hard sleeve and toward the intake.

Figure 4–15. The pressurized prime uses the jet siphon attachment found on some low-level strainers. Note the charged 1¾" line being utilized to prime the driver's side intake.

In order for this process to work, the pump operator must pump the jet siphon at an appropriate pump discharge pressure. If the jet siphon is underpumped, the water will not move through the jet siphon at a sufficient speed to create the necessary venturi. Depending on the strainer utilized, the jet siphon should be pumped at a pressure between 150 psi and 200 psi to move the water at the appropriate velocity through the jet siphon. Equally important for the operator to remember is that there is air ahead of the water coming up the hard sleeve. Failure to expel the air in the hard sleeve prior to opening the intake valve will cause the entire operation to cease as the air enters the pump.

There are two primary methods for expelling the air in the hard sleeve prior to opening the intake valve. The simplest option is to open the bleeder valve on the intake and allow the air to be expelled, which is similar to dealing with any other pressurized source. The only issue with this option is that depending on the size of the bleeder valve, it may take a long time to exhaust all the air from the hard sleeve. In addition, if the bleeder valve is located toward the bottom of the intake valve, not all of the air will be exhausted from the system. This could cause a problem when the operator opens the intake valve.

Another option for expelling the air during a pressurized priming operation involves loosely connecting the hard sleeve to the intake valve. Here, pump operators thread the hard sleeve onto the intake valve until they can feel the connection tighten against the gasket. At this point, instead of tightening the connection with a mallet, they should instead loosen the connection about a quarter of a turn.[16] By definition, there is not an airtight seal between the intake and the hard sleeve. At this point when the line feeding the jet siphon is charged, the air in the hard sleeve will be expelled around the loose connection at the intake. Once the operator sees water spraying/dripping around the hard sleeve connection, the connection should be tightened with a mallet until the dripping stops. This signifies an airtight seal has been achieved, and the intake valve can be opened.

After expelling all the air from the hard sleeve, the intake valve on the pumper can be opened fully. The biggest advantage to this technique is that it allows for a seamless transition between tank water and the external static source. After the intake is fully open, the tank-to-pump valve should be shut completely, and the tank fill valve should be opened to refill the booster tank. At this point, the operator no longer needs to keep the discharge feeding the jet siphon open in order to maintain a prime. While it is not incorrect for the operator to leave the jet siphon line charged, a portion of the pump's capacity will be used to flow water through that line. This may not be an issue in some scenarios, but it could be in others. One of the advantages of the pressurized prime is that after the prime is achieved, the jet siphon can be reopened at any point to serve as a recirculation line without wasting water in the source.

If this process is used to establish the initial prime, it is critically important to remember that it utilizes water from within the onboard booster tank. This means that the pump operator is on a time limit from the moment the jet siphon is charged. Operators can easily run out of tank water performing this technique if they are not proficient with the skill. This usually occurs because the operator fails to pump the jet siphon at the appropriate pump pressure or does not open the bleeder valve completely. If this technique is used to achieve the prime during a long reach operation, the pump operator must recognize that a large amount of air must be expelled, and thus it will take more time. Depending on the apparatus booster tank size, it is possible to run out of water before a prime is achieved using this method in conjunction with a long reach operation. During long reach operations, it is recommended to utilize the *burp drafting* technique discussed later in this chapter.

Some fire departments have the initial attack engine draft from a dump tank during rural water operations. This is generally not a recommended practice for many reasons. One reason is that it can be difficult to transition from tank water to the water in the dump tank without interrupting the flow to the attack line. For departments that do operate this way, using the pressurized priming method will result in a seamless transition between the two sources without affecting the firefighters on the line. In order to be successful, the operator must have a sufficient amount of water in the booster tank and must exhaust all the air prior to opening the intake valve.

The biggest advantage to using this alternative drafting method becomes apparent when the operator is attempting to prime a secondary intake. The pressurized prime allows the engineer to expel the air in the hard sleeve of the secondary intake prior to opening the intake valve. This will allow for an increase in the flow capability without negatively affecting the original prime.

Burp drafting

Burp drafting is another extremely efficient "primerless" priming technique that can be utilized in the rural environment. When this method is employed, the pump operator is literally burping the air in the hard sleeve through the pump in order to achieve a prime. The pump operator can use this method quite successfully in scenarios where the primer fails to operate. It is also extremely advantageous in situations where a significant lift is present or during a long reach operation. This is because the operator does not need to be concerned about burning up a primer pump/motor. It is critically important that the pumper have an intake valve on the intake that is being primed, which will enable the process to be conducted quite smoothly. It is equally important for the apparatus to have water in its booster tank.[17]

This method works on the premise of rapidly recirculating water within the pump. The operator's first step during this process is to open the tank-to-pump valve fully. Next, the tank fill valve should be opened. At this point, water is being recirculated between the booster tank and the pump. Water continues to recirculate because the laws of physics state that matter cannot be created or destroyed. The recirculation of water within the pump proves that when the pump is discharging water, it naturally creates an area of negative pressure at the eye of the impeller.[18] This enables more water to enter the pump through the tank-to-pump line from the booster tank.

If the operator were to increase the speed at which the impeller is spinning, more water would be discharged from the pump through the tank fill line. This will result in a greater area of negative pressure being created at the eye of the impeller, and thus more water will travel from the booster tank into the pump to replace the water that has been discharged. After water is recirculating through the pump, the operator should increase the throttle until the master discharge gauge reads approximately 150 psi.[19] At this discharge pressure, water is being recirculated fast enough to create a sufficient area of negative pressure at the impeller eye to perform the burp draft.

With regard to fluid dynamics, we know that fluids are attracted to areas of negative pressure. In this instance, the fluids involved are water and air, and their attraction to areas of negative pressure is the basis for burp drafting. After recirculating water and throttling up to 150 psi, the pump operator should slowly open the intake valve until there is a drop in the master discharge reading.[20] Since the rapid recirculation of the water in the pump creates an area of negative pressure, some of the air in the hard sleeve is drawn into the pump.

The drop in the master discharge reading signifies to the pump operator that the intake valve has been opened sufficiently to allow some air into the pump.[21] It is critically important that the operator *stop* opening the intake valve at this point. By doing this, a small amount of air is allowed to enter the pump and will be carried to the discharge side by the water entering from the booster tank. Because the tank fill valve is open during this process, the air and water is sent through the tank fill line back to the booster tank. The air then leaves the system via the tank vent.

The pump operator will notice that the reading will fluctuate on the master discharge gauge after the initial drop in pressure. This indicates that air is being burped through the pump. It is critically important that the pump operator wait until the pressure reading levels back out to 150 psi before opening the intake valve further. If the valve is opened too quickly or before the pressure levels back out, the operator will introduce too much air into the pump, and the process will cease. Once the pressure reaches 150 psi again, the intake valve should continue to be opened slowly until either another drop is noticed or until the valve is fully opened. If another drop in pressure occurs, the operator should stop as previously outlined until the air is burped through the pump. This may happen several times before the intake valve is fully opened due to factors such as the specific intake being utilized, the number of hard sleeve sections present, and the amount of lift being overcome.[22] Pump operators must remember that the pump will do the work for them if they have patience.

After opening the intake valve fully, the pump operator should slowly close the tank-to-pump valve (fig. 4–16).[23] Shutting the tank-to-pump valve without a pressure drop on the master discharge gauge signifies that a prime has been achieved. For this reason, it is critical that the pump operator slowly close the tank-to-pump valve since there may still be air within the intake plumbing. If the tank-to-pump valve is shut too rapidly, the last bit of air may enter the pump too quickly, causing the process to stop. Remember, the pump can handle small doses of air, but allowing too much air into the pump is what causes a complete loss of prime.

After the prime is achieved from the outside source, water entering the pump will be discharged directly into the booster tank via the tank fill line. This would cause the tank to overfill if the operator initiated the burp draft with a full tank of water. This serves as an additional confirmation that the

Figure 4–16. When burp drafting, the intake valve should be opened fully before any attempts to gate back the tank-to-pump valve are made.

pump operator has achieved a prime. The pump operator will want to close the tank fill line soon after noticing water overflowing, however. Not only will this prevent water from being wasted, but it will also prevent the rig from potentially getting stuck if it is not on pavement (fig. 4–17).[24] Even so, it will be important for the operator to remember that water must be discharged in some fashion in order to maintain the prime. The operator can achieve this either by supplying the scene or by opening a discharge that recycles water back into the source. If the booster tank was not full prior to the start of this technique, the operator should allow the tank to refill before closing the tank fill valve.

The major advantage of burp drafting is that no water is lost performing this technique as outlined above. Unlike the pressurized prime, the operator does not have to be concerned with running out of water in the booster tank prior to achieving a prime. This is because all the water used for this method is recirculated between the pump and the water tank, and it gives the operator more leeway in troubleshooting any problems encountered. By performing the burp draft, the pumper will literally run out of diesel fuel before the booster tank runs out of water (fig. 4–18).[25]

In situations in which the drafting engine is also forced to serve as the attack pumper, the burp drafting technique should be avoided. One reason

Figure 4–17. After achieving a prime using the burp method, be sure to immediately shut the tank fill valve. Failure to do so could cause the tank to overflow and the apparatus to become stuck.

Figure 4–18. The major advantage of burp drafting is that no water is lost in the process.

for this is because the air being burped will be sent directly through the attack line that is flowing. This is problematic since it will cause an interruption of water at the nozzle, resulting in erratic nozzle reactions as the air is expelled through the line. The operator must also be mindful of the fact that this process only works if water is flowing. If the nozzle is shut, a prime will not be achieved with this method.[26] A better alternative in these scenarios is to perform the pressurized prime because it allows for a seamless transition between water sources.

Engineers must be mindful of the size of the tank fill plumbing on their particular pumpers when performing this drafting method since it will have a direct impact on how far the tank fill valve can be opened when performing a burp draft. Apparatus equipped with 2" tank fill lines will likely not be able to have the tank fill line opened all the way. This is because the pump will likely cavitate prior to reaching 150 psi when just recirculating water.[27] If an apparatus is equipped with a 2" fill line, the tank fill valve will have to be gated back so that 150 psi can be achieved. Pump operators must ensure they do not gate back too much because that ultimately restricts the recirculation flow. This in turn will impact how much negative pressure is developed at the eye of the impeller. If the apparatus has a larger fill line, operators should train with their apparatus to find the sweet spot and mark it on the tank fill lever.

Burp drafting front and rear intakes

Front and rear intakes provide the rural pump operator with several options when operating from a draft. Many operators avoid utilizing these intakes for drafting operations because of the restricted flow rates they provide.[28] While this is true, the operator must recognize that these intakes provide unparalleled positioning advantages compared to side intakes. When establishing a dump site, these intakes allow the dump site to be built either in front of or behind the apparatus, while simultaneously allowing a lane of travel to remain open for tankers to off-load (fig. 4–19). During fill site operations, front intakes allow the pumper to nose in toward the source and establish a rapid water supply (fig. 4–20). In order to maximize the flow rate from the apparatus, a *twin tube drafting* configuration should be established, which will be discussed later in this chapter.

Some operators avoid using front or rear intakes because they are under the impression that these intakes will be more difficult to prime. This may be true if an onboard mechanical primer is used when the department is running a wet pump. When the onboard mechanical primer is used, the primer essentially has to evacuate the water in the pump casing before the air in the front/rear intake plumbing is exhausted. This creates one or both of the following situations:

1. The operator may burn up the primer prior to achieving the prime.
2. The operator may witness several periods where it seems there is a prime only to lose it again.

Figure 4–19. After backing into this driveway, this pumper burp drafted through its front intake to achieve its prime. Note how the dump site configuration leaves a lane and a half for tankers to off-load.

Figure 4–20. Nosing down this boat ramp allowed this fill site pumper to rapidly establish a water supply by burp drafting the front intake.

In situations where the pump is completely dry, the onboard mechanical primer will achieve a prime rapidly and effectively. This is essentially because there is no water in the casing that needs to be evacuated first. It is absolutely *not recommended* that a pump operator drain the water in the pump before attempting to prime a front or rear intake using the mechanical primer, however. Not only does this waste time, but the operator also runs the risk that the primer will fail to operate. For these reasons, it is more effective to use the burp drafting technique to achieve a prime on these intakes.

The operating principles for burp drafting a front or rear intake are no different than those outlined above. The only difference is the point at which the operator may see a drop in discharge pressure indicating air is being burped through the pump. When burping a side intake, the pressure drop happens almost immediately after the intake valve is barely cracked open. On a front or rear intake, however, the operator may open the intake valve fully and not see the discharge pressure drop at all. This does not mean the intake is primed.

As previously stated, front and rear intakes contain lots of bends, which will result in more resistance than with side intakes. If the intake valve for the front or rear intake is fully opened but there has been no corresponding drop in discharge pressure, it should signify to the pump operator that it is easier for the pump to receive water through the tank-to-pump line. In order to successfully prime these intakes, the operator has to make the front or rear intake become the path of least resistance. This is accomplished by *slowly* gating back the tank-to-pump valve. By restricting the water flow from the booster tank, the front or rear intake plumbing becomes the new path of least resistance.

The operator should slowly gate back the tank-to-pump valve until the master discharge pressure reading drops and then should stop gating. This signifies that air is now being burped through the pump. When the operator sees the pressure level out, the tank-to-pump valve should be gated back a little further. This process should continue until another drop in pressure occurs or until the tank-to-pump valve is completely shut. Once the tank-to-pump valve is completely shut, the operator has achieved a prime through the front or rear intake.

Although it may seem that the process for priming front and rear intakes differs from burp drafting a side intake, the steps are actually the same. Regardless of the intake being primed, the operator must remember to fully open the intake valve before closing the tank-to-pump valve.[29] On a side intake, the air will begin to be burped through the pump with a minimal opening of the intake valve. When this occurs, the operator stops opening the valve until the pressure levels out and then continues the process. Once the valve is fully opened, the operator begins to shut the tank-to-pump valve. On a front or rear intake, the operator may fully open the valve and not see any drop in discharge pressure. Again, only after the intake valve is fully opened should the tank-to-pump valve begin to be shut. The only difference between the side intakes versus the front or rear intakes is when the drop in pressure occurs. With side intakes, it will occur in the beginning of the process with the cracking of the intake valve. With front or rear intakes, it usually will happen on the back end of the process with the closing of the tank-to-pump valve.

Multiple Intake Drafting

Thus far, we have discussed the major principles related to drafting operations, reviewed conventional priming, and discussed several alternative priming methods. In this section, we will discuss options for increasing the flow capacity and efficiency of a drafting operation. When applying these tactics in the field, the operator must have a thorough understanding of the drafting principles and techniques previously outlined.

As stated earlier, once the dynamic intake reading on the compound gauge reads 22 inHg, it signifies to the pump operator that the current drafting setup has reached maximum capacity. Once the drafting operation is in service, there is not much the pump operator can do to reduce the amount of lift present. The pump operator can reduce the friction loss in the hard sleeve for the given flow rate, however. This is accomplished by establishing a prime on an additional pumper intake (fig. 4–21).

When multiple intakes are used simultaneously during a drafting operation, the total flow rate is split between the intakes. For example, imagine the pump operator is utilizing the driver's side and passenger's side main intakes with the same amount of hard sleeve and same strainers attached for a drafting operation. In this scenario, the total flow rate would be split equally between the two intakes in the operation (fig. 4–22). This means that the friction loss in the hard sleeve attached to each intake is 25% of what it would be if the total flow rate were flowing through a single hard

Figure 4–21. Priming an additional intake is the most efficient way to increase the capacity of a drafting operation.

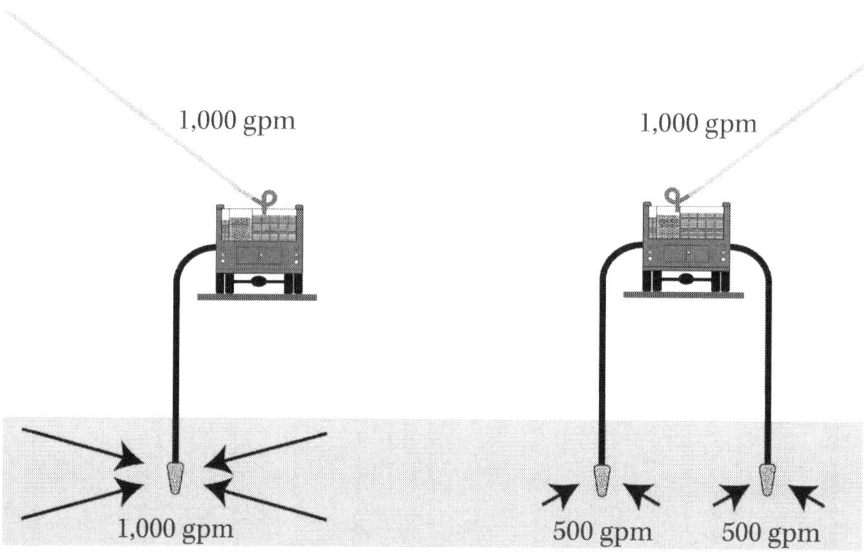

Figure 4–22. Because the pumper on the right has split the total flow rate between two intakes, the friction loss in each hard sleeve and intake is 25% of what it would be compared to the operation on the left.

sleeve and intake. The result is seen immediately in the form of a reduced dynamic intake reading on the compound gauge.[30] *NFPA 1900* outlines the maximum allowable friction loss values for hard sleeves and strainers at various flow rates.[31] Table 4–1 summarizes these friction loss values for 6" hard sleeve and shows the impact that drafting through multiple intakes has on friction loss.

It is important to note that it is not always realistic to use two identical intakes (driver's and passenger's main intakes) with the same amount of hard sleeve and strainers connected. In the real world, the operator may be using a front intake and a side intake. Or they may be using both the driver's and passenger's side intakes, but one intake has 20' of hard sleeve attached and the other has 30' attached. In these instances, the flow rate between each intake is not actually perfectly split between the two intakes. The pump operator should not be concerned with this difference, however, since the result will still be a dramatic reduction in the vacuum reading of the compound gauge. Table 4–2 highlights flow tests that were conducted on a 1,500 gpm pumper utilizing a twin tube drafting configuration with approximately 5.65' of lift present (fig. 4–23).

As evident from the data in table 4–2, multiple intake drafting operations can have a significant impact on the efficiency of an operation. When using a single intake, the pumper being tested was capable of flowing its rated pump capacity with a rather high vacuum reading. By performing a

Table 4–1. Effect of multiple intake drafting on friction loss within the hard sleeve and strainer assembly

Flow rate (gpm)	Friction loss in single 6"		Friction loss in twin 6"		Friction loss in triple 6"	
	Ft Water	inHg	Ft Water	inHg	Ft Water	inHg
1,000	3.4	3.0	0.9	0.8	0.4	0.4
1,250	5.2	4.7	1.3	1.1	0.6	0.5
1,500	7.6	6.7	1.9	1.7	0.9	0.8
1,750	10.4	9.3	2.6	2.3	1.1	0.9
2,000	—	—	3.4	3.0	1.5	1.3
2,250	—	—	4.3	3.8	2.0	1.8
2,500	—	—	5.2	4.7	2.3	2.0

Note: Cells with a dash (—) indicate values that are not realistic through a single hard sleeve configuration from a draft.
Source: National Fire Protection Association, *NFPA 1900: Standard for Aircraft Rescue and Firefighting Vehicles, Automotive Fire Apparatus, Wildland Fire Apparatus, and Automotive Ambulances, 2024* (Quincy, MA: NFPA, 2023).

Figure 4–23. Using a single intake, this pumper was capable of flowing 1,500 gpm with a dynamic intake reading of 18 inHg. After utilizing two intakes, the same 1,500 gpm was achieved with a dynamic intake reading of only 10 inHg. The flow was then increased to 2,000 gpm with a dynamic intake reading of only 13 inHg.

Table 4–2. These field tests demonstrate the effect that twin tube drafting operations have on the dynamic intake reading and achievable flow rate from the supply pumper.

Flow rate	Dynamic intake reading	Engine RPM
Single intake with 20' of 6" hard sleeve		
1,062 gpm	10.5 inHg	1,190 rpm
1,500 gpm	18 inHg	1,460 rpm
Twin tube through cab with drafting elbow: 20' of 6" on driver's intake, 30' of 6" on officer's intake		
1,500 gpm	10 inHg	1,470 rpm
2,000 gpm	13 inHg	1,620 rpm

twin tube drafting operation, the pumper was able to flow its rated capacity while creating the same amount of vacuum that the 1,000 gpm test yielded through a single intake. Even more impressive was the fact that the use of two intakes allowed the pumper to exceed its rated pump capacity by 500 gpm with a vacuum reading of only 13 inHg. Drafting through multiple intakes provides several benefits to the pump operator:

1. *The achievable flow rate is increased.* Once an additional intake is placed in service during a drafting operation, the operator will initially notice a reduction in the vacuum reading. This is because the total flow rate is being split between each intake and thus the friction loss is reduced. The only way that the dynamic intake reading would once again read the same as it did prior to placing the additional intake in service is for the flow rate to be increased.
2. *The workload on the apparatus is reduced.* If the desired flow rate is already being achieved with a single intake, performing a twin or triple tube drafting operation is still advantageous to the operator. When the total flow rate is split between multiple intakes, the vacuum required to maintain the desired flow rate is reduced. This means that the apparatus does not have to work as hard to provide the same volume of water. In many cases, if the flow rate remains the same after the additional intake is primed and opened, the pump operator will be able to decrease the revolutions per minute of the engine to maintain the discharge pressure required for the desired flow rate.
3. *A layer of redundancy is added to the operation.* This is especially true when dealing with dirty water sources. In the event that one of the strainers becomes clogged with debris, the operator is still able to flow water through the other intake while troubleshooting the problem. The result is a lower probability of an interruption in the water supply. It is also important to note that drafting through multiple intakes helps prevent the possibility of a strainer becoming clogged in the first place. When a twin or triple tube drafting configuration is employed, the velocity of the water moving in each hard sleeve is reduced dramatically. This is because the total flow rate is split and therefore the water in each separate hard sleeve does not need to travel as fast as it would if the total flow rate were moving through a single hard sleeve/intake. The result is that the water in the source around the strainer is not moving as fast to provide the same flow,

and therefore debris is less likely to be drawn toward the strainer. Table 4–3 compares the water velocity through 6" hard sleeve at different flow rates through twin and triple tube drafting configurations. Note that as the total area increases, the velocity through each hard sleeve decreases while maintaining a given flow rate.

Table 4–3. Effects of multiple intake drafting on water velocity through 6" hard sleeve at various flow rates

Flow rate (gpm)	Single 6" hard sleeve (ft/sec)	Twin 6" hard sleeves (ft/sec in each intake assembly)	Triple 6" hard sleeves (ft/sec in each intake assembly)
1,000	11.3	5.7	3.7
1,250	14.1	7.1	4.7
1,500	17.0	8.5	5.6
1,750	19.8	9.9	6.5
2,000	22.6	11.3	7.5
2,250	25.5	12.7	8.4
2,500	28.3	14.1	9.3

With regard to pump testing and certifications, *NFPA 1900* outlines the maximum number of intakes that may be utilized in order to achieve a given flow rate from a pumper. The standard also specifies the maximum amount of hard sleeve that can be used and the maximum amount of lift that may be present. Pump operators must remember that these requirements for pump testing may not reflect the scenarios presented in the real world. Rather, the pump operator should use the information provided by these standards as a baseline for understanding the conditions under which a pumper must be able to perform. Table 4–4 summarizes the requirements that must be met for different sized pumps to achieve their rated capacities.[32]

When comparing the rural fire scene to an annual pump test, it is important to remember that there are several distinct differences between the two:

- *Pump testing occurs in a controlled environment.* One of the biggest differences between multiple intake drafting for an annual pump test and for fireground operations is the uncontrolled and chaotic nature of the fireground.[33] If certain conditions are not right during a pump test, the test is stopped or postponed. During an emergency incident, however, the operator is expected to perform regardless of the conditions.

Table 4–4. Lift and intake configurations for NFPA 1900 pump certification tests

Rated capacity (gpm)	Maximum number of intakes with 20' of 6" hard sleeve attached	Maximum lift (ft)
1,000	1	10
1,250	1	10
1,500	2	10
1,750	2	8
2,000	2	6
2,250	3	6
2,500	3	6

- *All of the equipment is present prior to the start of the test.* The majority of pumping apparatus in the United States do not carry enough hard suction hose to be able to accomplish a twin or triple tube drafting configuration immediately. On the rural fireground, this requires these drafting setups to be constructed in segments as the equipment becomes available.[34] This means that the initial intake should be primed and flowing water before equipment is gathered and assembled for the second intake. When performing an annual pump test, the hard sleeve and strainers for both intakes are available and connected prior to the start of the test.
- *Primers are utilized for annual pump testing.* An apparatus automatically fails an annual pump test if the primer does not activate, if it does not achieve 22 inHg during the vacuum test, or if it does not prime the pump within the required time limit. On the rural fireground, however, it is not acceptable to abandon a drafting operation just because of a faulty primer. While alternative drafting methods are not considered acceptable means of priming the pump during a pump test, they are absolutely acceptable for use in the real world.
- *The primer is used to prime both intakes simultaneously during a pump test.* For pump testing purposes, the primer is the only device used to achieve a prime from a static source, meaning that this onboard device will be used to achieve the prime for all the intakes being utilized at the start of the test. In the field, one intake is usually primed and put into operation before the equipment becomes available to place a second or third intake into service. This means that the additional intakes are primed separately after the initial intake is in operation. In the field, the easiest way to achieve a prime on any additional

intakes is to perform one of the alternative drafting techniques previously discussed.

When performing a multiple intake drafting operation, it is critically important for the operator to start with a simple configuration and then expand it in segments as the equipment becomes available. Under no circumstances should the flow of water be delayed because equipment and/or resources are not available to prime the secondary or tertiary intake from the beginning of the operation. Pump operators should remember that we do not go from "zero to octopus." Instead, we build our supply system one "tentacle" at a time (fig. 4–24).[35]

Figure 4–24. Complex drafting configurations are built in segments as equipment becomes available.

The twin tube drafting configuration

As the name implies, a twin tube drafting configuration refers to utilizing two separate intakes during a drafting operation to achieve the desired flow rate (fig. 4–25). As outlined in table 4–4, this configuration is considered acceptable during an annual pump test for 1,500 gpm to 2,000 gpm pumpers to achieve their rated capacity. Nonetheless, this technique is not reserved only for the annual pump testing of these larger pumps.

The twin tube drafting configuration is considered the single most efficient way for increasing the flow capacity of a pumper during a drafting

Figure 4–25. The twin tube configuration is the most efficient way to increase the capability of any drafting operation.

operation.[36] When this technique is implemented, the pump operator will often see a dramatic drop in the dynamic intake reading at a given flow rate. This signifies a reduction in the workload of the pumper while the flow rate is maintained. After the twin tube drafting configuration is established, the operator will be able to either increase the flow rate or maintain the current flow rate with less strain on the apparatus.

It is particularly important to remember that this type of operation is built in stages. The first stage of the operation is to position the rig appropriately and establish the initial prime. For this reason, the operator should always position their supply pumper in a fashion that allows for the rapid establishment of a water supply while also allowing other apparatus access to the scene. In many instances, this may mean starting the operation with a front or rear intake.[37] While this may not be ideal from a flow perspective, it provides unmatched positioning advantages to the operator. In addition, by nosing in toward a water source, the pump operator needs less hard sleeve in order to reach the source.

Once the flow of water has been established through the initial intake, the operator can begin thinking about the establishment of a twin tube drafting operation. The next step in establishing a twin tube setup is to gather the necessary equipment for the operation, including additional hard sleeves and strainers.[38] With most fire departments, the pump operator will have to borrow this additional equipment from other apparatus on the scene. This is why it is advantageous for water shuttle apparatus to carry an abundance of hard sleeve.

Once the necessary hard sleeves and strainer are assembled, placed into the water source, and connected to the additional intake, the operator can begin priming the second intake. The goal here is to prime the secondary intake without interrupting the flow of water or disrupting the initial prime. The easiest way to do this is to perform a pressurized prime on the second intake.[39] This is because the operator is capable of expelling all the air in the secondary hard sleeve setup prior to opening the intake valve on the second intake. The result is an absolutely seamless transition from a single intake drafting operation to a twin tube drafting configuration, *as long as all the air is bled out.* Failure to exhaust all the air in the hard sleeve attached to the second intake will result in the immediate loss of prime on the initial intake, requiring the operator to reprime the initial intake. After expelling all the air and opening the second intake valve, the operator will see an instant drop in the dynamic intake reading.

While the pressurized prime is the easiest and most efficient way to achieve a twin tube drafting configuration, it may not always be an option to the pump operator. This is primarily because not all strainers are equipped with a jet siphon attachment that allows for this technique. This is especially true when using most barrel strainers. In these instances, the pump operator must make a slight adjustment to the method for obtaining a prime on the secondary intake. First, the operator will have to shut down the discharge(s) supplying water. Then, the operator should close the intake valve of the intake that was initially primed. By closing the first intake valve, the prime is held in the hard sleeve attached to the intake as long as there are no leaks present. This will allow the operator to prime the secondary intake either using the onboard primer or by performing a burp draft. After a prime is achieved on the secondary intake, the operator simply reopens the initial intake that was primed.[40]

Priming the second intake using the burp drafting method while simultaneously leaving the initial intake valve open *should absolutely be avoided.* This is because when the operator attempts to burp the second intake, the air that enters the pump will cause the prime to be lost on the initial intake. The result is a complete loss in the water supply, which will require the pump operator to reprime the first intake before anything else can be done. When the pressurized prime is not an applicable option, the operator should close the intake valve of the first intake prior to attempting to prime the second intake. This essentially enables the operator to capture the progress initially made. Once the second intake has been primed using the burp drafting method, the pump operator should reopen the intake valve of the initial intake. The flow of water can now be reestablished either to the fire scene or the fill site.

Although the twin tube drafting configuration is the most efficient way to increase the flow capabilities of a drafting operation, the pump operator must recognize when its use is warranted. Pump operators can make an educated guess as to whether this drafting setup will be necessary based on the role their pumper is serving on the rural fireground, whether as the dump site pumper or the fill site pumper. When serving as the dump site pumper, the pump operator should be focused on establishing a rapid and reliable water supply from the dump tank with the least amount of equipment possible. In many instances, this is accomplished using a front or rear intake.

The disadvantage of using front or rear intakes for drafting operations is that they may limit the flow rate to 500 gpm to 750 gpm depending on how the apparatus is plumbed.[41] Even so, this flow rate may be more than sufficient to supply the fire flow demands of the attack pumper on most rural fire scenes. The pump operator should pay close attention to the compound gauge to determine whether the intake configuration is sufficient to supply the flow demands of the fire scene. If the dynamic intake reading approaches 22 inHg, the operator should consider performing a twin tube drafting configuration at the dump site. This will allow the supply configuration to expand with the needs of the fire scene.

When operating as the fill site pumper, the pump operator is often faced with various obstacles that must be addressed. Some of these challenges include the lift that must be overcome, the setback from the source, and the amount of debris within the source. This is coupled with the fact that fill sites will likely be striving to fill mobile water shuttle apparatus at a rate of 1,000 gpm. For these reasons, the goal of the fill site should be to establish a twin tube drafting configuration as a standard practice (fig. 4–26). This configuration will enable the fill site pumper to consistently fill tenders at the desired 1,000 gpm fill rate. In addition, a twin tube setup will also reduce the total friction loss in each of the hard sleeve assemblies attached to the intakes. This will not only help the operator overcome more challenging lift scenarios but will also help to achieve the desired flow rate through long reach operations (fig. 4–27). The twin tube setup also provides

Figure 4–26. The fill site pumper should strive to achieve a twin tube drafting configuration as a standard practice.

Figure 4–27. The twin tube drafting configuration is extremely advantageous when it becomes necessary to perform a long reach operation.

a layer of redundancy to the fill operation in the event that one of the strainers becomes clogged with debris. This does not mean that the fill site cannot begin filling tenders until the twin tube drafting configuration is established. Rather, the fill site pumper should establish a rapid initial prime through a single intake to get the fill operation started. Only after this is accomplished should the operator begin establishing the twin tube configuration.

The triple tube drafting configuration

NFPA 1900 states that larger capacity pumps rated between 2,250 gpm and 2,500 gpm must be capable of doing so with no more than three intakes equipped with 6" hard sleeve.[42] This means that these pumps must perform a triple tube drafting configuration in order to achieve their rated capacity. In the field, the triple tube drafting configuration can be used successfully to increase the flow capability of pumps with capacities of less than 2,000 gpm (fig. 4–28).

While the triple tube drafting configuration does allow a pumper to flow large volumes of water while drafting, it also has disadvantages. First, these operations require a tremendous amount of hard sleeve. It is not unusual to utilize 100' of hard sleeve establishing this configuration (fig. 4–29). Successfully establishing these operations also requires a great deal of strenuous work, making it desirable to have several sets of hands. Often there is a shortage of personnel available on a rural fire scene to establish a drafting configuration such as this. One skilled pump operator can do it, but that individual will be exhausted afterwards. Establishing a prime on all three intakes will also require the operator to be extremely proficient with the alternative drafting methods discussed earlier in this chapter. Remember, pressurized prime is your best friend when it comes to multiple intake drafting.

Figure 4–28. By performing a triple tube drafting configuration, this 1,500 gpm setup was able to flow more than 2,000 gpm through a long reach operation. The limiting factor was the horsepower of the motor on the apparatus.

Figure 4–29. A tremendous amount of hard sleeve will be required in order to accomplish a triple tube setup.

For most municipal fire engines, the biggest limiting factor regarding the flow rate that can be achieved through a triple tube drafting configuration will be the horsepower of the motor on the apparatus. When this configuration is utilized, often the compound gauge will show an extremely low dynamic intake reading depending on how much lift is present. When attempting to max out the flow capability of the setup, however, the motor will likely reach its governed speed. When the operator observes a low dynamic intake reading, but the governed speed of the motor has been reached, it signifies that the horsepower of the motor is the limiting factor. In these instances, the limit has been reached because of the motor's inability to spin the pump fast enough to discharge the desired flow rate, not because of the intake configuration feeding the pump. This should be considered a good problem to have.

Operators must remember that discharge pressure is influenced by the amount of water leaving the pump. The higher the flow rate, the lower the discharge pressure. Thus when utilizing a triple tube setup to deliver large volumes of water, the pump discharge pressure will be correspondingly low. This does not mean that the revolutions per minute of the engine will also be low. During these operations, the operator should notice a low dynamic intake reading, high engine revolutions per minute, and low discharge pressures. This can be problematic if the apparatus is being used to relay pump or to build the necessary pressure to supply master stream nozzles. In these instances, the water supply officer should elect to perform a *high-volume series pumping operation*, which will be discussed in more detail in chapter 9.

While the triple tube drafting configuration allows a pumper to maximize the flow capacity, its realistic uses on the fireground may be limited. For this reason, the pump operator should consider the twin tube drafting configuration the best initial option for increasing the efficiency and capacity of a drafting operation. The twin tube setup yields impressive flow capacities while utilizing a realistic amount of equipment. For high-volume fires of long duration, the operator may consider implementing the triple tube drafting configuration for increased capability. Regardless, the operator must build these systems in stages as equipment becomes available.

Reaching the other side of the apparatus

In certain situations, the apparatus may be positioned with only one side facing the drafting source. This will make it easy to get the initial prime through one of the side intakes, but the operator may have to be creative in order to establish a twin or triple tube drafting configuration. This is where a front or rear intake comes in handy for placing an additional intake in service. When using a front or rear intake, the hard sleeve merely has to be directed toward whichever side of the apparatus the water source is on (fig. 4–30).

Engines not equipped with front or rear intakes will be forced to use the main intake on the opposite side of the pumper in order to establish a twin tube drafting configuration. A triple tube drafting configuration will almost

Figure 4–30. With the water source on one side of the apparatus, having a rear intake made it quite easy for this pumper to place a second intake in service.

always require use of at least one intake on the opposite side of the apparatus when the water source is located to the side of the apparatus. Exceptions to this are specialty apparatus that have dual 6" intakes on the driver's and/or passenger's side pump panel in addition to front and/or rear intakes (fig. 4–31).

In order to reach the opposite side of traditional municipal pumpers, some operators will run the hard sleeve underneath the rig (fig. 4–32). This

Figure 4–31. Having two steamers on the passenger's side of the apparatus makes it extremely easy for this pumper to perform a twin or triple tube drafting configuration.

Figure 4–32. Running hard sleeve under the apparatus in order to reach the other side of the rig should absolutely be avoided.

should absolutely be avoided because the heat emitted from the vehicle's exhaust could cause the hard sleeve to collapse. Not only will this irreparably destroy the hard sleeve, but it will also diminish the flow to the pumper.

The preferred method for reaching the other side of the rig is to run the hard sleeve through the cab of the apparatus (fig. 4–33). When pumping an apparatus with a top-mounted pump panel, the hard sleeve can be run across the pump panel walkway instead of through the cab if desired (fig. 4–34). While this technique may look extreme, it is the safest option for reaching the water located on the opposite side of the rig. This is because it protects the hard sleeve from the hot exhaust gases that it would be exposed to if it were routed underneath the apparatus.

A disadvantage to operating in this configuration is that it usually takes one full 10' section of hard sleeve to make the necessary 180° bend from the pump intake through the cab. If the department is using older nonflexible hard sleeve or hard sleeve rated for positive pressure, it may be impossible to bend the hard sleeve 180°. A 90° drafting elbow can be utilized to make connecting to the far side intake easier for the pump operator (fig. 4–35). The drafting elbow also takes the place of the additional 10' section of hard sleeve that has to be used when no elbow is present.

The operator should support the hard sleeve making the bend from the pump intake and through the cab (figs. 4–36 and 4–37). This is especially

Figure 4–33. Running hard sleeve through the cab of the apparatus is the safest way to reach the other side.

Figure 4–34. Hard sleeve can be run across the walkway of top-mounted pumpers in order to reach the other side of the apparatus.

Figure 4–35. A 90° drafting elbow can be used to help minimize the amount of hard sleeve needed to reach the other side of the apparatus.

true in the absence of a drafting elbow. Depending on the hard sleeve used by the department, the weight of the unsupported hard sleeve full of water could be enough to cause stress in the collars that fasten the couplings to the actual hard sleeve. This stress could cause separation in the hard sleeve

Figure 4–36. The weight of the water in the hard sleeve is causing it to drop dramatically, placing stress on the couplings.

Figure 4–37. The sleeve must be supported in the absence of a drafting elbow.

and couplings, allowing air to enter the pump and disrupt the prime. In addition, supporting the hard sleeve also helps prevent air pockets from forming inside the hard sleeve. This will make priming the second or third intake much easier.

Study Questions

1. At sea level, maximum lift is considered to be:
   a. 10'
   b. 15'
   c. 25'
   d. 33.9'
2. Which of the following is true regarding cavitation?
   a. Water temperature has no effect on cavitation.
   b. Cavitation occurs when the pressure on the intake side of the pump is lower than the vapor pressure of the water being drafted.
   c. Cavitation indicates that more water is available.
   d. Cavitation produces no noticeable sound.
3. Which of the following is a precursor to cavitation?
   a. Running away from water
   b. Reaching the governed speed of the motor
   c. Flowing more than 1,000 gpm
   d. Power filling
4. What does a dynamic intake reading of 22 inHg indicate to the pump operator?
   a. There is plenty of water remaining.
   b. The current setup is maximized and is close to cavitation.
   c. The current setup is minimized and is close to cavitation.
   d. Nothing; the gauge is reading incorrectly.
5. Burp drafting requires the operator to:
   a. Utilize the onboard primer
   b. Utilize a water eductor
   c. Have water in the booster tank
   d. Have an empty booster tank

6. Which of the following should the operator notice after performing a twin tube drafting operation?

   a. An increase in the dynamic intake reading
   b. No change in the dynamic intake reading
   c. Cavitation
   d. A decrease in the dynamic intake reading

7. Which of the following is an advantage of a multiple intake drafting operation?

   a. Increases the achievable flow rate through the drafting operation
   b. Reduces the workload on the apparatus
   c. Adds a layer of redundancy to the drafting operation
   d. All of the above

8. Which of the following is considered the single most efficient way to increase the capability of a drafting operation?

   a. Perform a twin tube drafting operation
   b. Draft through a rear intake
   c. Utilize the pressurized priming method
   d. Draft from a source with a lift of 25'

9. Which of the following engines should perform a twin tube drafting configuration as a standard practice?

   a. The attack engine
   b. The dump site engine
   c. The fill site engine
   d. All three of the above engines

10. Which of the following typically occurs when a triple tube drafting operation is implemented?

    a. Less than the rated capacity of the pump is achieved.
    b. The pump will exceed its rated capacity.
    c. The pump will cavitate.
    d. Nothing occurs since triple tube drafting is not possible.

5

DUMP SITE OPERATIONS

The rural environment inherently makes all aspects of firefighting more difficult. From a water supply perspective, rural fire companies typically do not have access to municipal water supplies. In the event that fire hydrants are present in the rural environment, they are usually sparsely located in the community and often provide less-than-desirable flow rates. In addition, many water sources may be too distant from the incident scene to make relay pumping a realistic option. These variables often require rural fire companies to employ some sort of water shuttle operation in order to deliver sufficient water to the fire scene.

A pump operator assigned to function as part of a rural water supply operation will most likely be assigned to operate at either the dump site or the fill site. The dump site is located in the vicinity of the fire scene and is responsible for supplying water to the fire attack pumper. This is usually where mobile water supply apparatus will off-load into dump tanks. The dump site pumper will typically draft from these tanks and send water to the fire scene. As the name implies, the fill site is where mobile water supply apparatus will go to be refilled by the fill site pumper. The most simplistic rural water supply operations will have both a fill site and a dump site in order to operate efficiently. More complex rural water operations may have multiple fill sites and dump sites with multiple pumpers at each location to deliver the required flow rate to the fire scene. The beauty of an efficient rural water supply operation is that it is able to expand or contract with the needs of the incident.

In this chapter, we will focus on strategies and tactics that are most efficient for dump site operations. Options commonly used for delivering water to the rural fire scene will be discussed and examined. In addition, the reader will be introduced to safe and efficient methods for establishing a dump site and maintaining the needed fire flow to the fire scene.

Note: The next two chapters will discuss operations related to mobile water supply apparatus. Using terminology compliant with the National Incident Management System (NIMS), these pieces of apparatus are referred to as "water tenders" or "tenders." Although not technically NIMS-compliant terminology, many areas of the United States refer to these pieces of apparatus as "tankers." Moving forward in this text, the terms "mobile water supply apparatus," "water tender," "tender," and "tanker" will be used interchangeably.

Delivering Water to the Rural Fire Scene

Many variables must be considered when establishing a water supply at the scene of a rural fire. Often the fire scene is located down a narrow driveway that makes accessing it with multiple pieces of apparatus quite difficult. The incident commander must also consider how many pieces of apparatus are responding to the incident in order to best determine where and how to deploy them. In this section, we will discuss options for delivering water to the fire scene safely and effectively.

Laying the supply line

For many rural fire departments in the United States, accessing the rural fire scene is the first challenge that must be overcome. The worst-case scenario is a fire that is located down a long, narrow driveway. Some fire departments will commit the attack apparatus down the driveway and build a dump site at the fire scene. This requires water shuttle apparatus to also commit down the driveway as well to off-load and then leave. Establishing a dump site in this manner should be avoided if at all possible. Not only does it reduce the efficiency of the shuttle operation, but there is also a much higher likelihood of an accident occurring with the tankers traveling up and down the narrow driveway.

The ideal fire scene configuration is for the dump site to be located near the mouth of the driveway, which will allow mobile water shuttle apparatus safe and easy access to the dump site. This will require a supply line to be laid from the mouth of the driveway to the attack engine operating at the fire scene. The first-arriving engine company must make the critical determination to either commit down the driveway without laying a supply line or to perform a forward lay from the entrance of the driveway. There are several variables that must be taken into account in order for the company officer to make this decision:

- *How many and what types of units are responding to the incident?* Certain organizations dispatch a specific number of pumpers and mobile water shuttle apparatus to a rural fire assignment. Other departments will send what they think they need based on the call and the area. Knowing how many and what types of apparatus are responding enables the company officer of the initial engine to determine whether or not they can spare the second-due engine laying the supply line up the driveway if the initial engine commits without laying one. If only three engines are responding to the fire, the initial engine may be forced to lay the supply line so that the second-due engine can serve as the dump site pumper. The third-due engine can then serve as the fill site pumper. It is also important to consider how long it will take for all these units to mobilize and arrive on scene.
- *How long is the driveway?* The length of the driveway will directly affect how much supply line must be laid and whether the initial pumper even has enough hose to complete the layout. If the initial engine commits to laying a supply line up the driveway without determining the distance, there is a potential for the engine to run out of supply line before reaching the scene. This will cause a significant delay in the establishment of a water supply, especially if the driveway is narrow. Having mapping software on the apparatus mobile data terminal (MDT) or through the dispatch center can be an invaluable resource in these situations. If these resources are not available, the operator can use the trip function of the vehicle's odometer to get a rough estimation of the driveway length. It may be helpful to remember that 0.1 mile equals 528', or roughly 500' of supply line.
- *Is there an standard operating procedure (SOP) or standard operating guidelines (SOG) that outlines what the initial engine should do?* Certain fire departments have written policies that dictate whether or not the initial engine company should lay up the driveway on a rural fire. These policies usually take into account the number of apparatus responding to the incident, as well as steps to be taken when dealing with longer driveways.

If resources and timing allow for it, many fire departments elect to have the first-arriving engine commit down the rural driveway without laying a supply line. The major advantage to this is that it allows the first-arriving engine to essentially serve as a scout for the incident. This will enable them to determine whether or not the next engine will have enough hose to complete the layout or if a multiple engine relay will be required. If it is determined that the next engine will have enough supply line to complete the layout, the second-arriving engine should perform a forward lay from the mouth of the driveway to the fire scene. After the second engine arrives, they should perform a booster backup operation to the first engine while a sustainable water supply is built at the entrance of the driveway.

The situation may arise where the second-arriving unit is a water shuttle apparatus. In these instances, a decision must be made whether or not to commit the tanker down the driveway in addition to the attack engine. If the apparatus is equipped with a pump, committing it to the fire scene could allow for an extended booster backup operation at the fire scene. In addition, if the tender is outfitted with a sufficient amount of supply hose, this apparatus can perform the forward lay to the scene. Once this apparatus is committed down the driveway, however, it should no longer be considered available for a shuttle operation. If it is determined that the tanker will commit down the driveway, the incident commander must request that an additional tender be added to the assignment. This will ensure that there is a sufficient number of tenders available for a shuttle operation.

Each fire department operating in the rural environment may have a slightly different approach to getting the initial supply line on the ground. The key is to remember the variables that must be taken into consideration prior to laying the line and to ensure that the supply line gets laid safely and efficiently. Once the supply line is on the ground, the dump site can be built around the entrance to the driveway. The sections that follow outline common configurations used to supply water to the rural fire scene.

Direct pumping operations

This method is arguably the simplest way to get water to the rural fire scene. It typically involves tenders equipped with a fire pump hooking to the supply line leading up the driveway and pumping their onboard tank water to the scene. For maximum efficiency, some departments place a *clappered siamese* at the end of the supply line leading to the fire scene to allow two tenders to hook up at the same time (fig. 5–1). Siamese appliances are designed

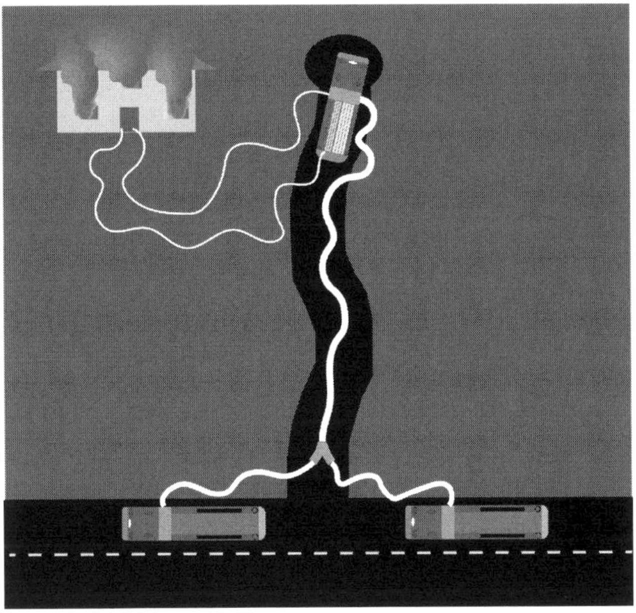

Figure 5–1. A direct pumping operation using a clappered siamese

to allow two separate supply lines to feed one single supply line. Most of these appliances contain a clapper valve that prevents the backflow of water. For rural water operations, it is crucial that any siamese appliance utilized has a built-in clapper valve. The principle here is that when the first tender is empty, water automatically flows from the second tender.

While this method is extremely simple to execute, it does have its disadvantages. Mobile water supply apparatus are most efficient when they are moving down the road. With this tactic, however, the apparatus will likely be stationary for a long time. If the initial handline at the fire scene is flowing 150 gpm, it will take at least 20 minutes to empty a 3,000-gallon tanker. In reality, it will usually take longer because the nozzle is unlikely to be fully opened for 20 straight minutes. This puts the tender sitting on scene for quite a long time.

Firefighters should remember that this is a critical tactic for the initial phases of an incident or small fire. Many fire departments use it as a replacement for dump tank operations, however. While this tactic has merit, firefighters must keep in mind that it is not the most efficient way to move larger volumes of water in the rural environment. Failure to recognize this will often result in a loss of water on the fire scene and an extreme delay in its reestablishment.

Nursing operations

Many fire departments use the term *nursing operations* to describe direct pumping operations, although there is a difference. While direct pumping operations involve the water shuttle apparatus pumping water to the fire scene, nursing operations are used when dealing with large capacity tankers. In a nursing operation, the large capacity tanker is essentially used like a dump tank on wheels. The dump site pumper uses hard sleeve to connect to the tank of the nurse tender and then drafts from it (fig. 5–2).

The major advantage this provides is that these large capacity tankers do not have to travel back and forth in a shuttle operation. This improves the safety of the operation because it keeps these larger, heavy apparatus from making repeated trips on rural roads. Once the large capacity tanker arrives on the scene, it positions to the side of the roadway, where the dump site pumper is able to connect hard sleeve to it (fig. 5–3). Once connected, it essentially creates an external tank-to-pump line from the dump site pumper to the nurse tanker. When additional tankers arrive, they simply pump their water into the nurse tender to refill it.

It is most efficient to off-load a tender using dump chutes, but nursing will likely reduce efficiency in the off-loading of tankers involved with the shuttle. This is due to the fact that tenders will have to pump their water into the nurse tender as opposed to simply dumping it into a tank. If the

Figure 5–2. In a nursing operation, the dump site pumper drafts water from the tank of a large capacity tanker. In this photo, the pumper is drafting water from the tank of this 5,000-gallon tender. Note the 3" lines on the ground. They are used to refill the nurse tender when smaller tenders arrive.

Figure 5–3. After the nurse tanker is appropriately positioned, hard sleeve is connected from the nurse tanker's tank to the dump site engine. The valve on the nurse tender will have a significant impact on the amount of water that is capable of being drafted. This valve limited the operation to 500 gpm.

tenders in the shuttle operation are equipped with larger tank-to-pump plumbing or dual tank to pump valves, however, they may be able to pump their tanks at the desired 1,000 gpm rate. Regardless, the major advantage to this tactic is an increase in safety when using large capacity water shuttle apparatus. Operators should remember that this tactic is not limited to use with larger capacity tenders only. Many fire departments successfully use this method for delivering water to fire scenes located on limited-access highways where it may not be feasible to set up dump tanks.

Dump tank operations

The use of portable tanks at the dump site allows for much higher flow rates to be achieved at the fire scene (fig. 5–4). The use of dump tanks allows tankers to position at the dump site, off-load, and then leave. It is important to remember that the *NFPA 1900* standard requires gravity-dump mobile water shuttle apparatus to be capable of off-loading at least 90% of their tank capacity at an average dump rate of 1,000 gpm.[1] In other words, a 2,500-gallon tender should be capable of off-loading 2,250 gallons of water in roughly 2 minutes and 15 seconds. This is a fraction of the time it would take for this same apparatus to be emptied during a direct pumping operation.

Figure 5–4. Dump tank operations allow much higher flow rates to be achieved at rural fire scenes.

The disadvantage of utilizing dump tanks is that it requires time to set them up properly. Certain fire departments disdain dump tank operations due to incidents where the fire scene has run out of water before the dump tanks were set up and filled. And time is not the only concern. Firefighters must be aware of the amount of space required to properly set up portable tanks at the dump site. If the dump site is not properly laid out from the beginning of the operation, there will be problems throughout the entire incident.

Many organizations refuse to utilize dump tanks simply because of the work involved with the operation. After all, it is a lot simpler to perform a direct pumping operation instead of setting up and tearing down multiple dump tanks. These fire departments must look at the operation from an efficiency standpoint instead. Although there is work involved with the setup of a dump tank operation, when this tactic is properly employed, much higher flow rates can be achieved at the fire scene. The question remains, "How do we rapidly establish water to the fire scene but also configure the dump site for efficiency?" The answer to this age-old question is a tactic called the *rural hitch*.

The rural hitch

The goal of any rural water supply operation is to rapidly establish water to the fire scene while also ensuring that the future flow demands are appropriately met. The rural hitch is a tactic used by many fire departments to establish a rapid initial water supply to the fire scene, while simultaneously building a dump site that is capable of larger flow rates. In essence, this tactic is a blend between a direct pumping operation and a dump tank operation. The success of this operation relies on the use of a clappered siamese attached to the end of the supply line leading to the fire scene (fig. 5–5). It should also be noted that this method requires at least the first tanker involved to be equipped with a fire pump.

Regardless of which apparatus lays the supply line, the rural hitch requires a clappered siamese to be attached to the supply line leading to the fire scene. When a tanker arrives on scene, it should pull past the driveway at least 100' and then connect a line to the siamese (fig. 5–6). Failure to pull past the driveway sufficiently will make it more difficult when it comes time to set the dump tanks. It is acceptable to utilize 2½" or 3" hose for the line between the tanker and the siamese. The major advantage to using an medium diameter hose (MDH) line for the initial phase of the operation is that it is lighter and easier to maneuver than an large diameter hose (LDH) line. Since the attack engine will be operating off its booster tank at the start of the operation, we know that it is not flowing more than 500 gpm initially. As we have previously discussed, this is a flow rate that is well within the capabilities of an MDH line. After making the appropriate connections and confirming the attack engine is ready for water, the operator

Figure 5–5. The clappered siamese is a critical piece of equipment for the rural hitch.

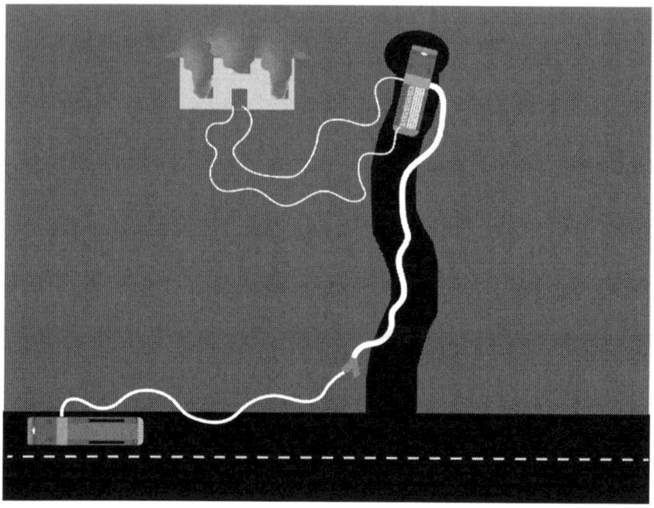

Figure 5–6. The first tanker arriving on scene should pull 100' past the driveway, connect to the siamese, and then initiate a direct pumping operation.

of the tanker can charge the line feeding the siamese. At this point, a simple direct pumping operation is being performed.

The next engine that arrives on scene will serve as the dump site pumper. This engine must decide whether to position in the driveway or out in the street in one lane of traffic. The advantage of either backing into or pulling into the driveway is that the pumper is out of the way of the tenders as they approach the dump site (fig. 5–7). The obvious disadvantage is that the pumper will most likely be blocking the entire driveway, making access for other apparatus impossible. By positioning the dump site pumper in the street, it is possible to keep the driveway open for other responding units. Regardless of where the dump site pumper is positioned, it is important to recognize that this apparatus will be drafting from tanks located either in front of or behind the apparatus (fig. 5–8).

Once the dump site pumper is properly positioned, the first priority should be to attach a supply line from a discharge to the unused inlet of the clappered siamese (fig. 5–9). This allows the dump site pumper to send their booster tank water to the fire scene in the event that the first tender runs out of water prior to establishing the dump site. This should only happen if the tender runs out of water in its tank. Ideally, a dump site pumper should ensure some water remains in the tank if the intent is to use an alternative drafting method. If the fire scene needs the tank water, however, the dump site pumper should not hesitate to send it. After these connections are made, the operator

Figure 5–7. Backing into the driveway keeps the dump site pumper out of the way of tenders attempting to off-load.

Figure 5–8. Keeping the dump site pumper in the roadway allows other apparatus potential access to the fire scene.

of the dump site pumper and the operator of the first tender should work together to start establishing the dump site. This will involve placing the dump tank in front of the dump site pumper and making the necessary hard sleeve and strainer connections. Once this is complete, the operation is prepared to transition from a pumping operation to a dumping operation.

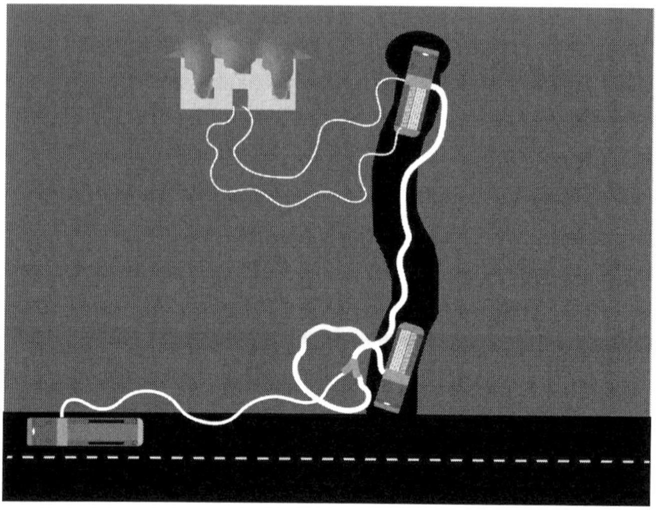

Figure 5–9. A supply line from the dump site pumper should immediately be connected to the open inlet on the siamese after positioning the dump site pumper.

The operator of the next tanker arguably has the most crucial decision to make in regard to the entire operation. The operator of the next tanker must decide whether to pump the onboard water or to dump it. This decision should be made after assessing whether the dump site is completely set up and determining how much water is remaining in the first tanker. If the dump tank is deployed, but the hard sleeve connections to the dump site pumper are not complete, the tanker operator should wait before dumping the water into the tank. This is because we do not want to have several thousand gallons of water sitting in the dump tank with an engine that is not prepared to draft it.

Equally important is the amount of water remaining in the tank of the first tanker. If the first tanker has more than a quarter of a tank left and the dump site is complete, the next tanker should dump its water into the dump tank. If the initial tanker has a quarter of a tank or less, however, the second tanker should connect to the first tanker's direct fill valve and refill the first tanker. This will prevent any connections on the siamese from having to be disconnected. Once the first tanker is refilled, the second tanker can leave for the fill site. The operator must remember that the intent of the rural hitch operation is to prevent any interruption in water supply to the attack engine. If the dump site is not set up and/or the first tanker is almost empty, a direct pumping operation should be continued.

Assuming the dump site is completed, and the first tanker has more than one-quarter of a tank left, the second tanker will off-load into the dump tank. As soon as the water level in the tank is above the low-level strainer, the dump site pumper should be working to achieve a prime. Once a prime has been achieved, the operator of the dump site pumper should immediately discharge water to the clappered siamese (fig. 5–10). After the dump site pumper charges its line feeding the siamese, the supply operation has effortlessly transitioned from a direct pumping operation to a dump tank operation. The beauty of this is that the attack engine should not notice a difference when the transition takes place.

Once the dump site pumper has taken over the supply operation, the initial tanker can shut down the discharge feeding the siamese. After disengaging its pump and disconnecting the hose from the pump panel, the initial tanker can leave to be refilled and participate in the shuttle operation. The 2½" or 3" line initially laid to the clappered siamese should be moved out of the way and remain connected to the siamese. This line will serve as a redundancy in the operation should a problem with the water supply arise. In the event that the dump site engine has some sort of mechanical failure, or the dump tank collapses, any tanker with a pump can hook to this line and supply water to the scene while the problem is corrected (fig. 5–11).

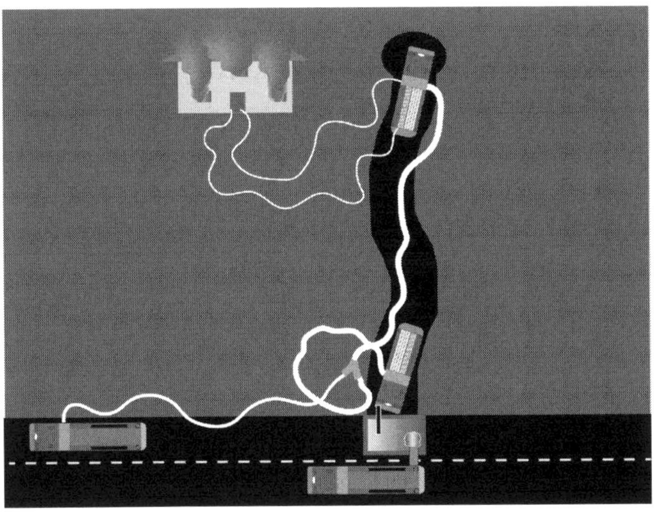

Figure 5–10. Once the dump site pumper has achieved their prime, they should immediately charge the supply line feeding the clappered siamese. This transitions the operation from a pumping to a dumping operation.

One of the major advantages to this tactic is that it allows the water supply operation to start out with a simple configuration and then expand in complexity based on the needs of the incident (fig. 5–12). Operators should understand that just because the dump tank is set up in front of the dump

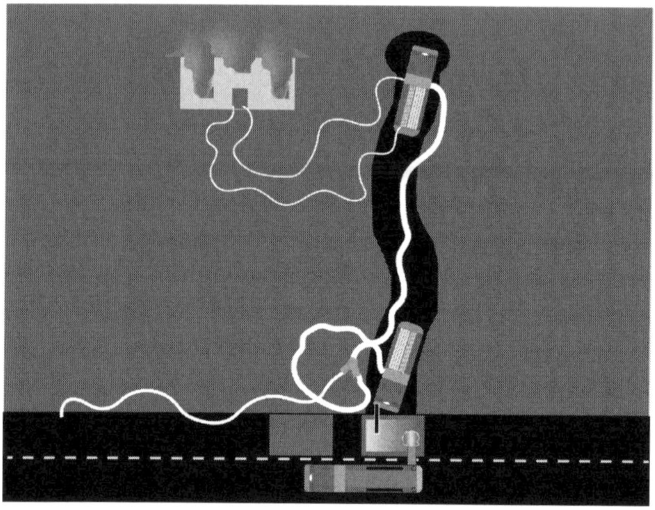

Figure 5–11. After the transition has taken place, the first tanker can leave to be refilled. The line they used to initiate the operation should remain connected to the siamese. Prior to the second tanker leaving, their dump tank should be removed from the apparatus.

Figure 5–12. The rural hitch allows the operation to seamlessly transition to a complex dumping operation.

site pumper, it does not mean that water must be dumped into it. It may be possible to manage small fires with a simple direct pumping operation from the initial tanker. If the fire flow demands increase, however, a dump site capable of meeting those needs will have already been established. The other beauty of this tactic is that it allows for a more efficient demobilization strategy at the end of the incident. Once the incident enters the overhaul phases, there may no longer be a need for a shuttle operation. At this point, the incident commander can assign one tanker to perform a direct pumping operation through the siamese while the remainder of the dump site is broken down.

Dump Site Considerations

Operators only have one chance to properly set up the dump site at a rural fire scene. This is why it is critically important that firefighters take the time to set the site up appropriately. Failure to do so will create issues for the entire incident. In this section, we will discuss factors the operator should consider when establishing a dump site to ensure the safest and most efficient operation possible.

Apparatus positioning and tank layout

Space is often quite limited when fighting fires in the rural setting. For this reason, operators must be conscious of how they position the dump site pumper and deploy tanks without impeding tanker access. As previously mentioned, for most fires in the rural environment, the dump site pumper can either be positioned in the driveway or in one lane of traffic. This decision will have to be made based on the variables present at each fire scene. For most fires in the rural environment, however, it is likely that the dump site pumper will be drafting from tanks located either in front of or behind the apparatus.

The first priority when establishing the dump site should be to make it easy and efficient for the tenders to off-load into the tanks. Always remember that the dump site is set up for the convenience of the tenders. Since many rural agencies utilize some sort of folding tank, it is important to position the tank where the straight edge of the tank faces the tenders that will be dumping from side chutes. If the hinged side of the tank faces tankers with side chutes, there is a likelihood of the tank collapsing inward as the water level gets low. This will make it harder for the tankers to get close enough to the tank to ensure all of their water is dumped inside of the tank (fig. 5–13). This problem will be exacerbated if the tanker has a short chute extension or no chute extension at all.

As the fire flow needs of the incident expand, so will the need for multiple dump tanks to be deployed at the dump site. Generally speaking, it is recommended that at least two dump tanks be deployed at every dump site in the rural environment. This will increase the amount of water that can be stored on scene while simultaneously increasing the efficiency of the shuttle operation. The initial tank placed will obviously be the primary tank that the dump site pumper drafts from to supply water to the scene. When tankers off-load into this primary tank, the force of the water entering the tank will entrain substantial amounts of air that will likely enter the intake of the dump site pumper (fig. 5–14). This will affect the prime of the dump site pumper and could possibly cause the rig to completely lose its draft. Placing a second tank on the ground enables each tanker to off-load into the second tank without concern of affecting the dump site pumper's draft. Water from the second tank will be transferred to the primary tank utilizing jet siphons.

The decision to place more than two dump tanks on the ground should be made based on the fire flow demands and whether or not there is a backlog of tankers waiting to dump. If tankers are waiting to dump, the efficiency

Figure 5–13. Avoid placing the hinged side of the dump tank facing the off-loading point for side-dumping tankers.

of the operation is being reduced (fig. 5–15). By placing another tank on the ground, tankers are able to spend less time sitting and more time moving in the shuttle operation. When utilizing multiple dump tanks, there are several dump tank layouts that fire departments elect to utilize based on the equipment available. The layouts discussed below are intended to give

Figure 5–14. The force of the water being dumped into the primary tank is causing air to enter the pump of the dump site engine. A second tank should be set up for use by the off-loading tankers to prevent this problem.

Figure 5–15. A backlog of tankers waiting to off-load is an indication that an additional dump tank should be put on the ground.

the reader a basic understanding of dump tank placement, but tanks do not have to be laid out exactly as they are described in the examples below. Firefighters should deploy their dump site based on the needs of their fire scene, not the words from a book.

The diamond layout. As the name implies, this dump site configuration has square or rectangular dump tanks rotated in a fashion that creates a diamond shape (fig. 5–16). This configuration is quite popular for departments that utilize tankers equipped only with rear dump chutes. These tankers are able to back into the "stall" that is created by the diamond shape of the tanks to off-load. If the tanker is also equipped with side chutes, this configuration will allow the apparatus the option of off-loading into two tanks simultaneously. The major disadvantage of this configuration is that it requires the largest footprint of all the configurations due to how the tanks must be rotated and how the tankers must be backed in toward the tanks (fig. 5–17).

The "offset L" layout. This option is extremely advantageous when the dump site pumper backs into the driveway, especially if the pumper backs in a little too far. The first tank should be set up in the driveway in front of

Figure 5–16. The diamond layout allows rear-dumping tenders the option to back up and off-load.

Figure 5–17. The diamond layout requires a substantial amount of space to accomplish. This may make it difficult for larger tenders to off-load.

the dump site pumper, flush with the roadway. This will keep both lanes of travel open for tankers initially. The second tank should be set up offset of the initial tank in one lane of traffic. If a third tank is placed, many firefighters will place it on the other side of the second tank where it completely blocks access to the primary tank. The major disadvantage here is that if the primary tank runs dry for some reason, tankers cannot easily access it in order to refill it. The better option is to put the third tank on the other side of the second tank to create more of an L-shaped layout (figs. 5–18 and 5–19).

The single-lane layout. As the name implies, every tank used in this layout is placed in one lane of traffic on the roadway. The dump site pumper can either be backed into the driveway or kept in the same lane of travel as the tanks. Either way, the tanks are set up in a straight line (fig. 5–20). When setting a dump site in this configuration, firefighters must be mindful to place the edge of each tank in line with one another. This will make it extremely easy for tankers with side chutes to pull alongside and off-load into whichever tank requires water. The other advantage of this configuration is that it can accommodate apparatus that are only equipped with rear dumps as well as apparatus equipped with side chutes (fig. 5–21). In scenarios where there is a mixture of tenders with dumping capabilities involved in the operation, dump site personnel should be mindful about saving the

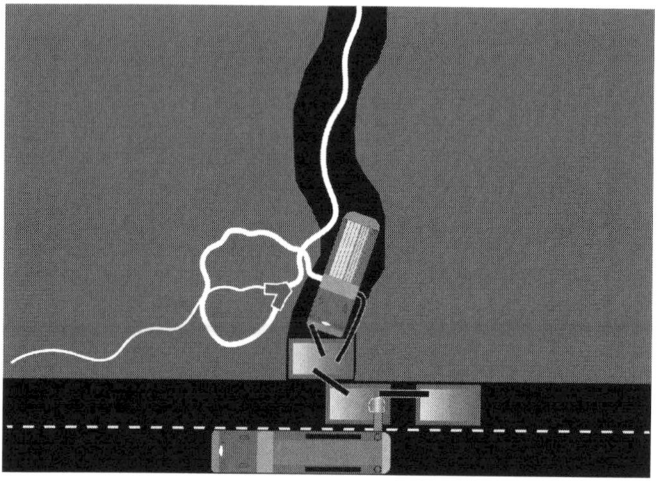

Figure 5–18. The offset L layout, with a side-dumping tanker off-loading in tank no. 2

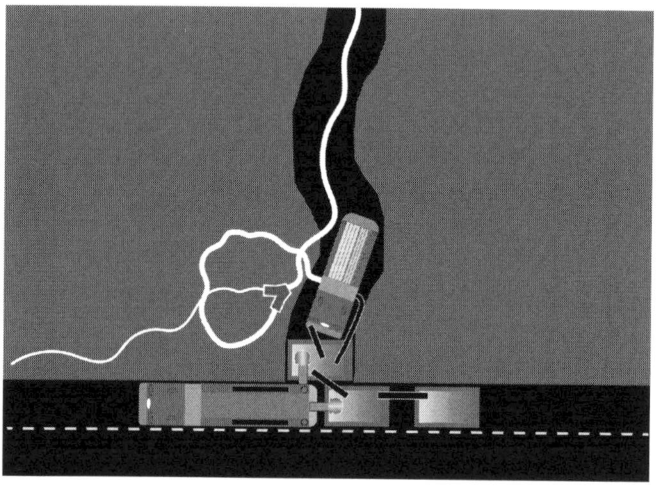

Figure 5–19. The offset L layout allows a side-dumping tanker to refill the primary tank in the event it runs dry.

furthest tank for off-loading by rear dump tenders. This will increase efficiency and keep downtime to a minimum.

The parking lot layout. This layout is a configuration used where space at the dump site is not an issue, typically in a parking lot. For this operation, the dump tanks can be set up in any of the previously outlined layouts. The

Figure 5–20. The single-lane layout keeps the dump tanks in one lane of traffic on the rural fire scene. Each tank is set up in a straight line down the street.

Figure 5–21. The single-lane configuration is capable of supporting both side-dumping and rear-dumping tenders.

major difference here is that there is enough space at the dump site for the dump site pumper to position alongside the tanks and easily draft from a side intake (fig. 5–22). There is also enough room for tenders to either pull alongside the tanks or back up to them to off-load. For fires in the rural setting, this option is rarely possible. This tactic may be utilized on larger

Figure 5–22. The parking lot layout is typically reserved for scenarios where there is enough space to position the pumper parallel to the dump tanks.

scale fires in urban settings, however, where rural water supply tactics are being implemented to supplement municipal water systems.

Water transfer between dump tanks

During a multiple dump tank operation, firefighters must ensure that there is a means for transferring water into the primary tank. The goal during any multitank operation is to ensure that the primary tank from which the dump site pumper is operating remains full at all times. Inefficient water transfer methods could cause the primary tank to run dry during the operation, which will negatively impact the fire scene. It is critically important for tanker operators and dump site personnel to understand that no tanker should off-load into a secondary or tertiary tank until a method for water transfer to the primary tank is established.

One of the simplest ways to transfer water between dump tanks is to run a hard sleeve between the drain chutes of two tanks. A rope or bungee cord is tied around the chute and hard sleeve to prevent leakage. Because the two tanks are connected via the drain chute, the level in both tanks will equalize and remain the same assuming they are on level ground. One of the disadvantages to this method is that it requires both tanks to be set up and connected together prior to water being dumped. Another major problem with this method is that as the water level is lowered, it is possible to have several hundred gallons of water in both tanks that the pumper cannot draft because it is spread over such a large area. It can be highly problematic when two large capacity tanks are tied together in this fashion.

The preferred method for transferring water between dump tanks is to utilize a jet siphon device. These devices are available as stand-alone units or come integrated in some low-level strainers and utilize the Venturi principle to transfer water between dump tanks. The jet siphon is connected to a section of hard sleeve run between two tanks. Hose is then run from the pump discharge of the dump site pumper to the inlet of the jet siphon. To transfer water between tanks, the dump site pumper simply opens the discharge feeding the jet siphon. It is critically important that the pump operator takes care to pump these devices at an appropriate discharge pressure to ensure a sufficient volume of water is transferred between the tanks. As a general guideline, it is recommended to pump a jet siphon at a pump discharge pressure of 150 psi.

Dump tank spacing has a significant impact on how efficiently the dump site is capable of operating. The ideal configuration places each tank roughly 6' to 10' apart. While this prevents the use of a single hard sleeve for jet siphoning water, it does provide other benefits to the dump site. One major benefit is that it reduces the possibility of injury or death by providing areas of safe haven for the dump site personnel. It also enables two tankers equipped with side dumps to potentially off-load at the same time, depending on the apparatus length and dump tank dimensions.

When dump tanks are spaced out in this fashion, firefighters must use a tactic known as *bridging* in order to transfer water between each tank. Bridging involves placing a ground ladder between two dump tanks and resting the hard sleeve used for the jet siphon operation on top of the ladder (fig. 5–23). The outlet of the transfer line can either rest on top of the ladder or be tucked underneath a rung so that it is directed downward into the primary tank. Only a few inches of the hard sleeve should be tucked underneath the rung of the ladder. This will prevent water from being siphoned backwards after the primary tank is full and the line feeding the jet siphon is shut down.

When operating from a single-lane tank layout, it is possible to have three dump tanks on the ground. Depending on how the dump site pumper is positioned, there will most likely be a tank positioned between the primary and tertiary dump tanks. Some fire departments will set up a leapfrog operation, in which they run a jet siphon line between tank three and tank two, and then from tank two to tank one. While this option certainly uses

Figure 5–23. During a bridging operation, a ground ladder is placed horizontally between dump tanks, and the hard sleeve of the transfer line is rested on top of the ladder.

less hard sleeve, it is not an overly efficient way of transferring water. This is because in order for water from the farthest tank to reach the primary tank, it must first be transferred to the middle tank and then transferred again into the primary tank.

Firefighters must remember that the goal of water transfer operations is to ensure that the primary tank remains full at all times. In a three-tank operation, there will still be a transfer line running between tank two and the primary tank. The transfer line from tank three should be bridged over the top of tank two and should dump directly into the primary tank, however. This configuration ensures that water always goes directly to the primary tank, where it is most needed (fig. 5–24). Generally speaking, extension ladders work best for bridging over the top of the secondary tank. If the tanks are spaced far enough apart that an extension ladder will not fully reach from tank one to tank three, a straight ladder can be used to overcome the remaining distance (fig. 5–25).

For most basic rural dump site operations, the dump site pumper will be the apparatus powering each of the jet siphons. In a three-tank system, this will usually involve at least two jet siphon lines. The operator may elect to attach the lines feeding each of these jet siphons to individual discharges on the pumper. While this is a valid option, it requires the operator to open each individual discharge every time water needs to be transferred between

Figure 5–24. The transfer line from the third tank is bridged over the second tank and is able to dump directly into the primary tank.

Figure 5–25. The transfer line from tank three required the use of two ladders in order to reach the primary tank.

tanks. In addition, the pump operator may not be able to see each of the tank levels from the pump panel, making it difficult to know exactly when to open each discharge.

If the apparatus is equipped with a 2½" discharge on the front bumper, an easier option is to attach a gated wye to this discharge and supply each jet siphon from it. This enables the pump operator to simply keep the front discharge valve open during the entire operation and to control the transfer of water from the gated wye (fig. 5–26). Operating in this fashion allows the operator to see the water level in each tank during the operation. If the apparatus is not equipped with a front discharge, another option is to run an MDH line from a discharge on the pump panel and then attach the wye to the end of the line. This will accomplish the same goal as using a front discharge to power this operation.

Situations may occur where there is a sufficient amount of water in each of the dump tanks at the dump site, but it cannot be transferred fast enough into the primary tank. In these situations, the operator should elect to put multiple transfer lines in service between each tank (fig. 5–27). This will allow water to be transferred between each tank at roughly twice the rate. The water used to power the jet siphons is known as *motive*. As more jet siphons are placed in service by the dump site pumper, the motive to power them uses more of the pump's rated capacity. If the dump site pumper is

Figure 5–26. By placing a gated wye on the front discharge of the dump site pumper, a firefighter is able to watch the transfer of water to ensure the primary tank does not run dry.

Figure 5–27. Two transfer lines from the second tank were used at this dump site to ensure the primary tank was refilled faster than it was depleted.

powering two jet siphons, the operator should consider employing a twin tube drafting configuration. This will ensure that the pumper is capable of supplying the needed fire flow to the scene while also having enough pump capacity left over to power the jet siphons. If more than two jet siphons must be placed in service, the water supply officer should consider assigning a dedicated transfer pumper to power the jet siphons.

Dedicated transfer pumpers

As the fire flow demands increase at the rural fire scene, the dump site pumper may lack pump capacity to provide the needed fire flow to the attack engine and power the jet siphons between tanks. In these instances, a dedicated transfer pumper should be utilized to power the jet siphons. This will enable the dump site pumper to use all of its pump capacity to provide water to the fire scene while the transfer engine focuses on keeping the primary tank full. The dedicated transfer pumper should establish its own prime out of the primary dump tank in order to power each of the jet siphon devices. Establishing the draft of the transfer engine from the primary tank decreases the possibility of the dump site pumper or the transfer pumper losing a prime.

After the dedicated transfer engine has established its prime, each of the lines powering the jet siphons should be disconnected from the dump site pumper and connected to discharges on the dedicated transfer engine.

One of the biggest challenges to successfully placing the dedicated transfer engine in service is getting the transfer engine close enough to the primary tank in order to establish a draft. This is where proper tank placement and layout becomes critical. If the dump site pumper has backed into the driveway and a single lane layout has been established, it is important to ensure that the tanks are laid out in sequential order down the street: primary tank, then secondary tank, and finally tertiary tank. This will allow the dedicated transfer pumper to simply nose right up to the primary tank to establish a prime utilizing the least amount of hard sleeve (figs. 5–28 and 5–29).

If the single lane layout is utilized and the dump site pumper is kept out on the street, it will likely make it more difficult for the transfer engine to reach the primary tank. These scenarios will require the transfer pumper to use multiple sections of hard sleeve in order to reach the primary tank (fig. 5–30). Another option is for the transfer pumper to utilize a device known as a *TurboDraft* in order to reach the water in the primary tank. TurboDraft operations will be discussed in more detail in the next chapter.

Pump operators should remember that dedicated transfer pumpers do not necessarily have to be municipal pumping apparatus. These operations can successfully be put into service using mini-pumpers or even portable pumps. One of the advantages of using these smaller units is that they may

Figure 5–28. The pumper on the right is serving as the dump site pumper, while the pumper on the left is serving as the dedicated transfer pumper. Note how both engines are drafting from the primary tank.

be able to be positioned off the road in order to draft from the primary tank. If smaller pumping units are used, it will be necessary to ensure that they have a sufficient pump capacity to transfer the necessary volume of water between tanks at a pump discharge pressure of 150 psi.

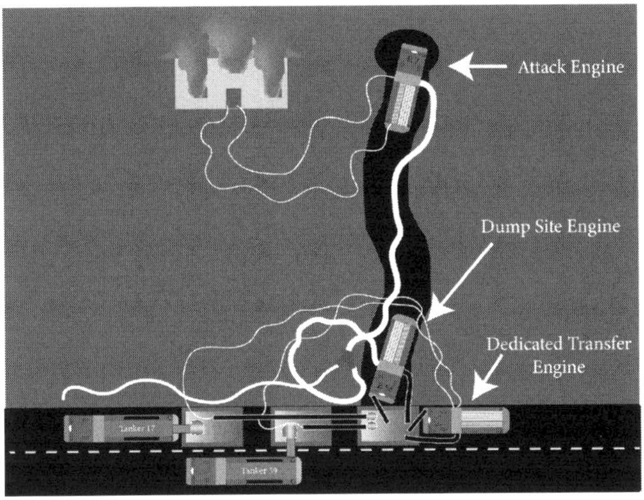

Figure 5–29. One of the major advantages of backing the dump site engine into the driveway is that it allows a dedicated transfer engine easy access to the primary tank should it be needed.

Figure 5–30. The dedicated transfer pumper may be forced to perform a long reach if it cannot easily access the primary dump tank.

Study Questions

1. Which of the following is true regarding the rural hitch?
   a. It requires the attack engine to draft from the primary dump tank.
   b. It is most successful when the dump site pumps into a gated wye to feed the fire scene.
   c. It is intended to transition the rural water supply from a direct pumping operation to a dumping operation.
   d. All of the above are true.

2. What should the second-arriving tanker do if the dump site is built but the first tanker is at one-quarter of a tank of water?
   a. Dump into the first tank.
   b. Pump the water to the dump site pumper.
   c. Deploy a second dump tank.
   d. Pump the water into the tank of the first tanker.

3. Where will the dump tanks be positioned in relation to the dump site pumper on the majority of rural fire scenes?
   a. On either side of the dump site pumper
   b. In front of or behind the dump site pumper
   c. Always where the operator can view the tank
   d. Always at the rear of the apparatus

4. Dump sites should be set up for the convenience of the __________.
   a. Attack engine
   b. Dump site pumper
   c. Tankers
   d. Dedicated transfer pumper

5. Which of the following dump site layouts is most advantageous for organizations that only have rear-dumping tankers?
   a. Diamond
   b. Offset L
   c. Single lane
   d. Parking lot

6. What is the generally accepted pump discharge pressure for jet siphons?

    a. 50 psi
    b. 100 psi
    c. 150 psi
    d. 200 psi

7. Which of the following pieces of equipment is utilized for a bridging operation?

    a. Ground ladder
    b. Hose clamp
    c. Clapper siamese
    d. Gated wye

8. Which of the following is an indication that an additional dump tank should be set up at the dump site?

    a. The dump tanks currently in service are running dry.
    b. More than two tankers are involved with the operation.
    c. There is a backlog of tankers waiting to off-load.
    d. It is never necessary to drop more than one dump tank.

9. If more than __________ jet siphons are placed in service at the dump site, a dedicated transfer pumper should be implemented at the dump site.

    a. One
    b. Two
    c. Three
    d. Four

10. According to the scenario in question 9, where should the dedicated transfer pumper obtain its water supply from in order to power each of the jet siphons at the dump site?

    a. The fill site
    b. Its booster tank
    c. The primary dump tank
    d. The secondary dump tank

6

FILL SITE OPERATIONS

The establishment of a fill site capable of maintaining water flow that is adequate to meet the demands of the incident is critical to the success of any rural water supply operation. To ensure this is accomplished in a timely manner, an engine company should be assigned to serve as a fill site pumper from the initial dispatch. In addition, the closest water source to the fire scene should be identified and utilized as the initial fill site for the incident. It is important to recognize that this may not always be a fire hydrant. More often than not, the closest water source is a static body of water.

Operating from static water sources will pose some of the most challenging scenarios that a pump operator may encounter. In some cases, it may be difficult or impossible for the operator to utilize the water within these sources. Many of these challenges can be overcome by preplanning each rural water source, however. This enables the fire department to develop strategies for accessing the water from the viable sources identified. Two variables have the biggest impact on the usability of the source and must be considered during preplanning:

- *Accessibility of the source.* There are many water sources in the rural environment that might be adequate for use in terms of volume but do not allow a safe means of access for the pumpers. Sometimes the ground surrounding the source may not be able to support the weight of the apparatus, or access to the source may be blocked by some sort of physical barrier that makes it difficult to reach the water. In some instances, the apparatus may not have enough hard sleeve or other equipment to reach the water. A lack of equipment could easily make the most bountiful water source useless for firefighting operations.

- *Adequacy of the source.* This refers to whether or not there is a sufficient volume of water contained within the source to utilize. The fire department must identify whether the body of water is subject to any seasonal fluctuations that may render it unusable. Some sources may provide sufficient water only during certain times of the year, where others may be bountiful year-round. There may also be sources that could provide a sufficient volume of water but are too shallow for purposes of drafting. Yet other sources may contain so much debris that drafting from them would force the operator to deal with clogged strainers throughout the entire incident.

Overcoming Access Issues

Obtaining access to a particular body of water is one of the most familiar challenges that a fire department faces. Many rural fire companies forgo a water source simply because they cannot reach it. In this section, we will examine some options available to the fire department for accessing more challenging water sources. Some of these options require installation of specific features such as dry hydrants, while others require the department to invest in particular pieces of equipment. Each individual fire department must evaluate which particular solution works best in their operating environment.

The long reach

As stated earlier, the long reach is an option for accessing distant water sources by connecting several sections of hard sleeve together (fig. 6–1). In order for this tactic to work, the fire department must ensure that there is not an excessive amount of lift present that could diminish the flow rate. The biggest disadvantage to this tactic is that it requires several sections of hard sleeve. Depending on the amount of hard sleeve transported on the pumper, the engine operator may have to wait for additional sections of hard sleeve to arrive on other apparatus before a prime can be established.

If a department routinely operates from distant water sources, it would be advisable for them to carry 40' to 50' of hard sleeve on the pumper. Even if this hard sleeve is not used for long reach operations, most dump site configurations can quickly employ 50' of hard sleeve. Because a long reach

Figure 6–1. Long reach operations will require the operator to use more than 20' of hard sleeve to reach the source.

operation requires several sections of hard sleeve to be connected together, it is critically important that operators ensure that all sections have gaskets and that the connections are airtight. Failure to do so will introduce air into the system and will prevent a prime from being achieved.[1]

A long reach operation may need to be implemented even if the pumper can position close to the water source. This could be necessary if the water source has a shallow bank, and the strainer has to be pushed further out into the source before there is a sufficient depth for the drafting operation. Sources that contain a lot of vegetation and/or debris along their banks could also require a long reach operation in order to push the strainer out past the area where the debris will cause clogging issues (fig. 6–2).

Boat launches/paved accesses

Water sources equipped with a boat launch enable a fire department pumper to nose in directly toward a source that may otherwise be inaccessible. Apparatus equipped with front intakes can rapidly obtain a water supply when positioning on a boat ramp. Nevertheless, the operator should be prepared to establish a twin tube drafting configuration to maximize the drafting capability (fig. 6–3). Depending on how the ramp is constructed, it may have a concrete pad underneath the water that extends for several feet past the shoreline. This may help prevent the strainer from clogging.

Figure 6–2. A long reach operation was performed in order to get the strainer past the point where the vegetation would cause blockage issues.

Figure 6–3. Boat launches make nosing in toward a water source extremely easy. Remember to perform a twin tube configuration in order to maximize the efficiency of the operation.

Typically, the biggest disadvantage of a boat launch is that it may be difficult to train with during peak recreational seasons. Blocking a boat launch during a training evolution will likely not be well received by the public. Even so, the fire department should not hesitate to use these access points during an actual emergency incident. Operators must also recognize that most boat ramps are not wide enough for the rig to position sideways for a side intake. When tasked with using a boat launch, the operator should be prepared to nose in toward the source. If dealing with a rear-mounted pump or a rig with a rear intake, it may be necessary to back the apparatus down the ramp (fig. 6–4). Regardless of whether the apparatus is nosed in or backed down the ramp, operators must be extremely cautious and take their time approaching the water to prevent a catastrophe. In addition, they must remember to place wheel chocks after exiting the apparatus.

Dry hydrants

Many rural communities install dry hydrants near static water sources where there are access challenges. Dry hydrants serve as excellent options for accessing distant water sources due to the fact that they reduce the total amount of hard sleeve needed to reach the source. They are typically composed of PVC piping that is run underground and into the water source at a sufficient depth. The end of the plumbing should have a strainer attached to prevent debris from entering the pump.

Operators should remember that dry hydrants can naturally become clogged from vegetation and wildlife.[2] This is why many fire departments annually flush and flow test dry hydrants in their response district to ensure they are in working order. It is critically important that the pump operator

Figure 6–4. Backing down the ramp enabled this pumper to utilize its rear intake to achieve the initial prime. Later, a twin tube configuration was established.

flush a dry hydrant prior to use for any fire incident, and certainly before attempting to pull a prime from one. The best option for flushing a dry hydrant requires the pump operator to connect a 1¾" or 2½" hose to the dry hydrant connection and pump a portion of the tank water into the dry hydrant. It is recommended that the pump pressure not exceed 80 psi during the flushing process. The operator should watch the water source for bubbles to emerge, signifying that the plumbing and strainer are sufficiently blown out (fig. 6–5). The operator should not pump more than one-half of the booster tank water into the dry hydrant during the flushing process. This will ensure water is still available to perform a burp draft if the primer fails to operate.

Pressurizing the dry hydrant also provides the operator with the benefit of identifying any cracks or leaks that may be present in the dry hydrant elbow or connection. Depending on the severity of the crack or leak, the operator may be able to overcome it with some duct tape or cellophane wrap, or the dry hydrant may have to be taken out of service. The key concern to remember when using dry hydrants is to ensure that they are flushed and do not have any cracks or leaks before committing them to use.

Once the dry hydrant has been sufficiently flushed, the operator must begin the process of connecting the hard sleeve to the dry hydrant. Many pump operators will attempt to do this using a single 10' section of hard suction hose. This should be avoided because it makes it extremely difficult

Figure 6–5. When backflushing a dry hydrant, the operator should look for bubbles to emerge from the strainer, indicating it has been sufficiently flushed.

to appropriately line up the side intake with the dry hydrant connection, especially when the pump operator is alone. In addition, once the single 10' section of hard sleeve is full of water, the weight of the water will cause the hard sleeve to droop in the middle if it is not supported. This places a significant amount of stress on the hard sleeve couplings, as well as the dry hydrant itself. If a single 10' section is used to connect to a dry hydrant, it is highly recommended that the operator support it in the middle in some fashion (fig. 6–6).

When the operator is tasked with connecting to a dry hydrant single-handedly, it is recommended that two 10' sections of hard sleeve be used. The operator should position just short of the dry hydrant initially and perform the flushing procedure previously outlined. Once the flushing has been completed, the operator should take two 10' sections of hard sleeve, connect them together, and then connect to the dry hydrant. This 20' section of hose creates a longer "lever arm" and therefore is easier to bend. After the connection is made to the dry hydrant, the operator should bend the entire assembly toward the direction of the pumper and lay it on the ground. The pumper can then be pulled forward until the side intake is in line with the middle of the first hard sleeve section. This will enable the end of the hard sleeve to be easily maneuvered and connected to the pump intake. In addition, the pumper will also have a smaller footprint in the roadway (fig. 6–7).

Figure 6–6. Be sure to support the center of the hard sleeve if a single 10' section is used to connect to the dry hydrant.

Figure 6–7. Using 20' of hard suction hose not only makes it easier to make the connection to a dry hydrant, but it also allows the pumper to maintain a smaller footprint.

A potential disadvantage of dry hydrants is that they generally result in lower flow rates compared to traditional drafting operations. Generally speaking, operators can expect dry hydrants to flow anywhere from 750 gpm to 800 gpm depending on how they are installed. Another reason it benefits fire departments to annually flush and flow test their dry hydrants is that it provides a realistic idea of the flow rate that can be expected from these sources. If the flow from the dry hydrant is less than desirable, it does not mean that the operator is out of options. A twin tube drafting operation can still be performed after the initial prime is achieved utilizing the dry hydrant (fig. 6–8). This will enable the operator to quickly get water flowing using the dry hydrant and then expand the system to meet the desired flow rate with an additional intake. The limiting factor here will be whether or not there is sufficient hard sleeve available to overcome the distance to the source.

Figure 6–8. This pumper established a twin tube drafting configuration after using the dry hydrant for the initial prime. This dramatically increased the flow rate from the pumper.

Portable pumps

Many rural fire departments utilize portable pumps to access water from distant sources. While these units vary in pump capacity, it is critically important to remember that the majority of these pumps are considered high-volume, low-pressure pumps. This means that they are designed to flow their rated capacity at a low pump discharge pressure. For example, it is common for these pumps to have a capacity of 500 gpm at a pump discharge pressure of 25 psi to 35 psi.

Using a single portable pump for fill operations typically results in long fill times because these pumps are designed to flow their rated capacities at low discharge pressures. Since higher flow rates result in exponentially higher friction loss values within the supply hose, the pressure produced by the portable pump will likely be insufficient to support the desired fill rate. For these reasons, if a portable pump is utilized for a fill operation, it is highly recommended that large diameter hose (LDH) lines be connected to the discharge ports of the portable pump. Using LDH lines will enable the pump to deliver more of its rated capacity at a lower pump pressure due to the reduced friction loss within the hose.

The efficiency of a portable pump operation can be increased when multiple portable pumps are used in conjunction to support the fill operation. In this configuration, two or three separate portable pumps are placed at the edge of the water source, each with LDH lines extended from their discharge ports. The hose from each portable pump should then be connected into either a siamese, "tri-mese," or LDH manifold. On the outlet side of the appliance, another LDH line is connected that can be used to fill tankers directly, feed a fill site engine, or discharge directly into a dump tank (fig. 6–9). Although this configuration takes work and practice, it does allow for impressive flow rates from remote water sources.

One of the major disadvantages of using portable pumps is their size and weight, and they can be difficult for a single firefighter to carry. The majority of these units are powered by an auxiliary gasoline or diesel motor, which means that they must be properly fueled, charged, and routinely operated in order to be reliable on the fireground.[3] Failure to maintain these units prior to the incident will result in them becoming a useless paperweight on scene.

Water eductors

An option for rapidly accessing a remote rural water source with minimal personnel is to utilize a *water eductor*, also known as a TurboDraft (fig. 6–10), as mentioned in chapter 5. These devices operate on the Venturi principle and function in a similar fashion to a jet siphon.[4] The major difference between a TurboDraft and a traditional jet siphon is the size of the device

Figure 6–9. Multiple portable pumps being utilized to supply a fill site engine

Figure 6–10. A TurboDraft being used to access a distant water source

and the output flow rate. These devices can be utilized in several ways on the fireground for rural water supply operations. It is critically important, however, to understand that these tools are primarily designed to allow rural pump operators access to water sources they otherwise would not be able to reach (fig. 6–11). Although a TurboDraft operation may not yield the same flow rates as a traditional drafting operation, the major advantage of these devices is how rapidly they can be deployed in order to reach distant water sources.

Unlike traditional drafting operations that use hard suction hose, conventional TurboDraft operations utilize soft suction LDH to deliver water to the pumper. This enables the devices to be put into service quite rapidly. Using a TurboDraft in a conventional configuration requires a 2½" or 3" line to be connected to the inlet of the device and the other end connected to a pump discharge. Next, the operator must connect a minimum of 5" LDH line to the outlet side of the device. The other end of the 5" hose line should then be connected to an intake on the pumper. After these connections have been made, the TurboDraft can be placed into the water source. In most situations, the operator should ensure that the device is placed in the water with the strainer facing upward to prevent silt and debris from clogging the strainer and pump.[5]

Figure 6–11. The water source pictured here was 75' from where the apparatus could reach it. In addition, it was 20' below the apparatus. The TurboDraft enabled the pumper to access the water and flow 500 gpm.

Options for placing the device into the water include the following:

1. *Tying it to a ladder and sliding it into the source.*[6] This option is great for dealing with water sources that have steep banks and are relatively deep. It is also useful when dealing with fast-moving bodies of water that have the potential to cause the device to flip upside down and clog the strainer of the device. This method should be avoided when dealing with sources with shallow banks because the ladder tends to dig into the muck, which may prevent the device from being sufficiently submerged.
2. *Attaching a float to the device.*[7] Attaching a float to the TurboDraft is by far the easiest way to deploy it into a water source (fig. 6–12). One of the most useful floats is an empty foam bucket, which can be attached to the anchor point on the device with a carabiner. The operator simply throws or pushes the TurboDraft into the water once the float is attached and the hoses are connected. To ensure the device is sufficiently submerged, the operator should see the foam bucket start to be pulled underwater.
3. *Utilize an aerial device to lower the TurboDraft into the water.* While this option is not as common, it is useful when dealing with sources that have fast-moving water, steep or otherwise treacherous banks,

Figure 6–12. Attaching a float to the TurboDraft is the easiest way to deploy the device.

and other hazards present (fig. 6–13). Obviously, this is only a viable tactic if the aerial apparatus can safely position near the water source. In these situations, the aerial is simply being used as a high point for lowering the device into the water. It is *not* being used to anchor the TurboDraft in place against the water current. Under no circumstances should a tortional or lateral load be placed on the aerial device when employing this tactic. The tension of the hose lines connecting the TurboDraft and the pumper should be what keeps the device from floating downstream.

4. *Get in the water and place the TurboDraft.* In some instances, the source remains shallow for a good distance from the shoreline. The distance to deeper water may prove too far for a firefighter to throw or push the TurboDraft. In such circumstances the operator may have to get into the water to deploy the device. Companies that routinely operate in the rural environment and face this scenario should carry a set of waders on each pumper. A float should still be attached to the TurboDraft in these instances.

After the TurboDraft has been placed in the water source, and the operator has confirmed all the connections have been made, it is time to begin

Figure 6–13. Deploying a TurboDraft using an aerial device

the operation. With the pump in gear, the operator should open the tank-to-pump valve fully. Next, the operator should fully open the discharge valve connected to the TurboDraft. As the 2½" or 3" line feeding the device is charged, the throttle should be increased to 200 psi on the individual discharge feeding the TurboDraft. This is critically important because failure to pump the line at this pressure will prevent a sufficient Venturi effect from being created at the device.[8] This will result in a lower return flow rate from the TurboDraft.

Almost immediately after charging the line feeding the water eductor and throttling up to the appropriate pressure, the operator will see the 5" supply line start filling with water. Prior to opening the intake valve, the pump operator must remember to bleed the air in the supply line. After bleeding all the air from the supply line, the operator should fully open the intake and then slowly close the tank-to-pump valve.[9] At this point, the pump operator is pumping water from the external static source (fig. 6–14). In the current configuration, a portion of the water coming from the TurboDraft is being used to send water back to the device to keep the process going. This is essentially a giant recirculation line.

It is critically important for the pump operator to understand that TurboDraft operations require a portion of the booster tank water in order to initiate the process. For this reason, the operator must ensure that there is water in the booster tank prior to initiating the process and that no time is wasted bleeding the air and opening the intake because it could cause

Figure 6–14. A pumper accessing a static water source via a TurboDraft operation

the tank to run dry during the process.[10] If there is no water in the tank, or the tank runs dry while initiating the water eductor operation, a water supply will not be established through this operation. The only option available in these circumstances is to refill the booster tank prior to reinitiating the TurboDraft operation.

Once the supply of water has been established from the TurboDraft operation, the operator should immediately refill the booster tank. Unlike a traditional positive pressure water source, the intake gauge during a TurboDraft operation will read approximately 0 psi. This is because we are not technically dealing with a closed system, and thus the 5" supply line is not truly pressurized. Instead, the water is directed up the supply line due to the Venturi effect caused by the water eductor. This means that if the operator is not careful and opens a discharge valve too quickly or starts flowing more water than the TurboDraft can provide, the supply line will collapse. This will immediately cause all water flow from the pump to cease. The solution to this problem is for the operator to reopen the tank-to-pump valve to jump-start the flow of water through the system.[11] This is why it is critically important for the pump operator to ensure that the booster tank is full at all times during the operation. Once the flow of water is reestablished, the operator should ensure the discharge flow rate is not exceeding the capability of the TurboDraft and then close the tank-to-pump valve. Finally, the operator will once again top off the booster tank.

Throughout the entire TurboDraft operation, the discharge valve feeding the device must remain fully open, and the pressure on that discharge must remain at 200 psi. If either one of these conditions is not met, the efficiency of the operation will diminish or the supply line will collapse. During a conventional TurboDraft operation, the return flow rate from the device is dependent on the distance between the pumper and the device, the diameter of the hose used, and the lift present. As a general rule, the minimum hose diameter used for these operations should be 5" LDH, and the maximum distance between the source and the pumper should be 100'.[12] Using smaller supply hose or laying more than 100' of hose will result in less-than-ideal return flow rates during conventional TurboDraft operations.

The operator can expect a return flow rate ratio from the TurboDraft of roughly 2.5:1 when using 100' of 5" hose with minimal lift present. As mentioned previously, the water used to power the device is known as the motive. At a pump discharge pressure of 200 psi, the motive is slightly more than 200 gpm. With 100' of 5" hose between the pumper and the device, the usable return flow rate from the TurboDraft is approximately 550 gpm.[13] As the distance between the pumper and the TurboDraft is reduced, the return flow ratio will increase. When deploying the TurboDraft, the operator must be mindful to use only the amount of 5" hose needed to deploy the eductor. In other words, the pump operator should not use 100' of 5" hose when 75' will suffice. Using more hose than necessary causes two problems for the operator:

1. The increased hose length causes more friction loss in the supply line and therefore reduces the available return flow from the eductor.
2. Longer hose lengths will likely need to be flaked out in some fashion to get rid of the extra hose on the ground. As water fills the supply line, kinks will likely form in the hose that will drastically diminish the return flow to the pumper.

If a conventional TurboDraft operation is being utilized at a fill site to fill water shuttle apparatus, it will likely be unable to provide the desired 1,000 gpm fill rate for the operation. Even with 50' of 5" hose separating the apparatus and the TurboDraft, the operator can only realistically expect a 700 gpm return flow rate.[14] The operator has three choices available when seeking to increase the flow capability of a TurboDraft operation: 1) Increase the hose diameter; 2) Perform a tandem TurboDraft operation; or 3) Use a TurboDraft to fill a dump tank, then draft from it.

Increase the hose diameter. The laws governing friction loss state that when larger hose is utilized, friction loss is reduced tremendously. In relation to TurboDraft operations, an uncomplicated way to increase the return flow rate from the device is to utilize 6" LDH instead of 5" LDH. Tests performed show that just by switching out 50' of 5" hose with 50' of 6" hose, the usable return flow rate from the TurboDraft increased from 700 gpm to approximately 950 gpm (fig. 6–15).[15] This allows a single conventional TurboDraft operation to efficiently fill tankers close to the desired fill rate. It is also possible to use 6" soft suction hose to maximize flow from fire hydrants, which will be discussed in the next chapter.

Occasionally there is a need to make up a distance of 10' to 20' between the end of the 5" soft sleeve and the pumper's intake. In these situations, if the operator does not have 6" soft suction available and does not want to use 5" because of the friction loss characteristics, the operator can use 6" hard sleeve to make up the distance. Not only does this make up the difference in distance, but it also reduces the friction loss due to the fact that it is larger in diameter.[16] In addition, it also prevents a kink from forming at the pumper's intake.

Perform a tandem TurboDraft operation. This is one of the simplest ways to increase the flow capacity of a TurboDraft operation. Essentially, the

Figure 6–15. Six-inch soft sleeve being utilized to increase the capability of a TurboDraft operation

operator places a second TurboDraft into operation off of the same engine supplying the first unit. This will effectively double the return flow rate back to the pumper (fig. 6–16). All that is needed to accomplish this is a second water eductor and a duplicate hose layout hooked to the pumper. The beauty of this tactic is that the process for putting the second TurboDraft into service is no different than putting the first unit into service. It is important to remember that the operator must ensure that pump discharge pressures on both of the discharges feeding the TurboDraft remain at 200 psi for the entirety of the operation.[17] Although the usable return flow rate from the tandem water eductors will be in the 1,000 gpm range, in reality, the pumper will be moving close to 1,400 gpm, including the motive for each device. For this reason, it is critically important to ensure that the pumper performing this tactic has a large pump capacity and a sufficiently powered motor.

Use a TurboDraft to fill a dump tank, then draft from it. This tactic is hands-down the most complex to perform and requires the most equipment and personnel. Nonetheless, it truly enables operators to utilize water sources that might otherwise be considered inaccessible (figs. 6–17 and 6–18). One of the biggest advantages to this tactic is that the operator does not have to worry about collapsing the supply line during the operation. Instead of connecting the 5" return line to the pumper's intake, it is used instead to fill a dump tank located near the pumper.[18] After discharging a portion of

Figure 6–16. A tandem TurboDraft operation

the tank water to begin filling the dump tank, the operator should achieve a traditional prime and continue the process.

The major advantage here is that the pumper is capable of drafting its rated pump capacity out of the dump tank.[19] The operator just needs to be careful not to outpace the return rate from the TurboDraft. Generally speaking, for distances up to 100', two TurboDrafts should be utilized for every 1,000 gpm consistently flowed. When being used at a fill site and with an adequate time interval between tenders, a single TurboDraft may be sufficient. The pumper will draw the tank level down during the filling process and then refill the tank while it waits for the next water shuttle apparatus to arrive.

Although it is not required, to achieve peak efficiency, two pumping apparatus should be utilized for this tactic (fig. 6–19). This allows one rig to focus on powering the TurboDrafts while the other can focus on filling water shuttle apparatus. When the water source is more than 100' from the location where the pumper can make access, this TurboDraft configuration should be considered. In addition, this is an extremely useful tactic for large-scale fill site operations where there is insufficient room to position multiple fill site pumpers at an abundant source. By using this method in these scenarios, the third or fourth fill site pumper is essentially able to

Figure 6–17. The engine on the left is powering two TurboDrafts that are 150' from the pumper and down a 30' hill leading to the Mississippi River. The TurboDrafts are being used to fill a dump tank from which both engines are drafting. The engine on the right was able to maintain a flow rate of 1,000 gpm.

Figure 6–18. Two TurboDrafts being used to access water from the Mississippi River

Figure 6–19. This tactic works best when two pumpers are utilized—one to power the TurboDrafts, and the other to flow water from the tank.

move the water source to a location where they can make use of it.[20] Another application for this tactic is when the water source being utilized is particularly dirty or silty. When this tactic is used in conjunction with these sources, it allows the debris from the water source to settle to the bottom of the dump tank. This helps prevent trash from entering the supply pumper by creating a simple filtration process.

Determining the Adequacy of the Source

Prior to committing to any water source, the pump operator must be confident that the source has a sufficient volume of water to support the needs of the rural water supply operation. This requires the fire department to routinely evaluate the rural water sources in their district over the course of the year to determine the usability of the source. Several methods exist to determine how much water is contained within a particular source based on whether it is flowing or static. When dealing with a square or rectangular static source, the following equation can be utilized to determine how much water is contained within the source (Vol_{gal}):

$$Vol_{gal} = L \times W \times D \times 7.5$$

where

L = Length of the source in feet
W = Width of the source in feet
D = Average depth of the source in feet
7.5 = Number of gallons within one cubic foot

When dealing with a circular static source, the following equation for volume expressed in gallons should be used:

$$Vol_{gal} = \pi \times r^2 \times D \times 7.5$$

where

$\pi = 3.14$
r = Radius of the circular source in feet
D = Average depth of the source in feet
7.5 = Number of gallons within one cubic foot

If the operator is dealing with a flowing river or stream, a different equation must be utilized. This equation takes into account the velocity of the water flowing in the source to determine the gallon per minute flow rate of the source. The equation for flow rate in gallons per minute (Q_{gpm}) is given as follows:[21]

$$Q_{gpm} = W \times D \times V \times 7.5$$

where

W = Width in feet
D = Average depth in feet
V = Velocity of the water in feet per minute
7.5 = Number of gallons in one cubic foot of water

While the calculations for the static sources are fairly straightforward, determining the flow rate of a flowing source is more involved. When calculating the adequacy of a flowing source, the hardest variable to determine is the velocity of the water flow. To calculate this in the field, pick two points on the shoreline 10' apart. Next, place a buoyant object in the source at the upstream point and time how long it takes to travel past the downstream point. Since we have identified a predetermined travel distance of 10', we can determine a conversion factor that makes these calculations easier to perform in the field. For example, if it took the object one second to travel between the 10' markers, the velocity would be 10'/sec. Converting this to feet per minute would be calculated as follows:

$$\frac{10 \text{ feet}}{1 \cancel{\text{sec}}} \times \frac{60 \cancel{\text{sec}}}{1 \text{ min}} = \frac{600 \text{ feet}}{1 \text{ min}} = 600 \text{ ft/min}$$

As long as 10' is always used as the distance between the markers on shore, we know that any amount of time measured in seconds will have to be divided by 600 in order to determine the velocity in feet per minute. Therefore, the following conversion can be used (see Appendix B for precalculated source volume for static and flowing water sources):

$$\textit{Velocity} \text{ (ft/min)} = \frac{600}{\textit{Time to travel } 10 \textit{ feet} \text{ (sec)}}$$

Shallow Water Sources

Fire departments often encounter shallow water sources and discount them as viable options simply because of the water depth. Before completely eliminating the source as an option, the operator must first determine if the shallow source is flowing or static. Shallow water that is flowing may give the operator more flexibility with regard to the water supply operation. While these sources may not have enough depth to perform a traditional drafting operation, they may provide a usable volume of water for filling tankers. In these situations, the pump operator has several options for utilizing the water from these sources.

The traditional solution to dealing with a shallow flowing water source is to perform a damming operation. These operations involve using a tarp wrapped around a ground ladder to create a dam, which is then placed into the water source (fig. 6–20). The tarp should be wrapped around the ladder until a tail roughly 3' long remains. As the dam is placed into the water, firefighters must make sure that the tail faces upstream. This serves to prevent underflow currents from going around the bottom of the dam. Depending on how quickly the water is moving, rocks or other heavy objects may need to be placed on top of the tail to prevent water from flowing underneath the dam. In addition, as the water level rises, there will be a significant amount of force applied to the dam, which could cause a failure. Hand tools or pickets may need to be driven into the ground or creek bed to help prevent the dam from collapsing.

The width of the flowing source will dictate what length of ladder must be utilized. Sources that are especially wide may require the use of multiple ground ladders and tarps. Alternatively, narrow channels could make it more difficult to use a ground ladder for the dam structure. In these instances, the operator should consider using a backboard for the structure of the dam instead of a ladder.

It is highly unlikely that any damming operation will raise the water level in the source to the desired 4' recommended depth for drafting with a barrel strainer. This means that box strainers or low-level strainers will likely be used for these operations. Box strainers draw water from the top and are ideal in these instances because they reduce the likelihood of silt or other debris entering the pump. Even so, they require the water to be sufficiently deep to prevent whirlpools from forming. When using a low-level

Figure 6–20. This damming operation sufficiently raised the water level of this source to allow for a large flow operation. Note the bracing on the back side of the dam using common fire department hand tools.

strainer in these situations, the operator must ensure that the bottom of the source is fairly clean and not full of debris. If the bottom of the source poses a significant clogging concern, the operator should elect to perform an alternative method for drafting from this type of source.

The pump operator should recognize that any debris floating in the water upstream will eventually make it down to the dam. This could cause a situation over time where a large amount of trash collects around the dam and therefore near the strainers. If this debris is allowed to collect around the dam, there is a high likelihood of the strainers clogging. A handline equipped with a fog nozzle can be deployed and sprayed around the strainers to prevent debris floating on top of the water from getting drawn toward the strainers (fig. 6–21). This stream can also be used to help break up any whirlpools that may form around the strainers. The firefighter operating the nozzle must maintain a fog pattern and not plunge a straight stream into the water around the strainers. If this occurs, air will be entrained into the pump, and the stream will likely stir up sediment at the bottom of the source.

Another option for dealing with shallow water sources is to utilize TurboDraft equipment. This option works well with both static and flowing bodies of water. When a shallow static body of water is being considered for use, operators must first determine if there is enough water in the source

Figure 6–21. This fog nozzle is being used to break up any whirlpools that form and to keep debris away from the strainer.

to make it worth the effort. Next, the operator must determine how shallow the source is and whether or not the bottom is free of silt and debris. This will impact which water supply tactic will be employed with the TurboDraft.

For shallow sources that are 10"–12" deep, the TurboDraft can be placed into the source right side up as usual.[22] If the water source has a dirty or silty bottom, orienting the TurboDraft in this configuration will help prevent any clogging issues from occurring. If the water source is less than 10" deep, however, the TurboDraft will entrain air into the system and be rendered useless. A float should still be attached to the device if it is being used in a shallow source to help break up any whirlpools that form.

Another option for overcoming sources that have debris on the bottom and are at least 10" to 12" in depth is to use the TurboDraft as a conventional strainer. In this configuration, the TurboDraft is connected directly to a hard sleeve and is used as the strainer through which a traditional drafting operation occurs. This allows the device to physically sit on the bottom of the source while simultaneously minimizing the potential for silt and debris to clog the strainer (fig. 6–22). This option is particularly useful when dealing with shallow creeks and streams that are flowing.

Scenarios may occur where the water source is extremely shallow but has a clean rock bottom and is flowing. In these situations, it is acceptable

Figure 6–22. A TurboDraft being used as a conventional box strainer in a shallow, silty water source. The float is being used to break up whirlpools.

to flip the TurboDraft upside down in order to use the water in the source (fig. 6–23). Using the TurboDraft in this configuration will allow the operator to access the water in sources as shallow as 4" to 6" in depth.[23] Prior to employing this tactic, the pump operator must ensure that the bottom of the source is clean and free of silt and debris. Failure to do so could cause major damage to the pump on the apparatus. If the operator has any doubts as to the presence of silt and debris at the bottom of the water source, the TurboDraft should be used to fill a dump tank, and then drafting can be conducted from the dump tank. Operating in this fashion helps to filter the water from the source as debris is allowed to fall to the bottom of the dump tank. This should help prevent any clogging issues.

Fire Hydrants as Fill Sites

Although hydrant operations will be discussed in much greater detail in the next chapter, there are a few points that should be made regarding their use as fill sites in the rural environment. Many fire departments in the United States utilize fire hydrants as the sole source of water for fill

Figure 6–23. Placing the TurboDraft upside down in a natural water source is only acceptable if the source has a clean, rocky bottom.

operations. Some organizations will even have their tenders drive past perfectly good static water sources in order to be refilled at a fire hydrant. While the reasons for this vary, operators must recognize that it is not acceptable to drive past water in order to get water. One of the main pillars of successful rural water supply operations is to utilize the closest possible water source to the fire scene, even if it fills tenders below the desired 1,000 gpm fill rate. If a fire hydrant is determined to be the closest water source to the fire scene, steps must be taken to ensure that the full potential of the hydrant can be achieved.

Filling directly off the hydrant

Some organizations will simply connect a single medium diameter hose (MDH) line to the hydrant, connect the other end to the direct fill of the tanker, and use hydrant pressure alone to refill the apparatus. Although this does refill the shuttle apparatus, it is typically highly inefficient. Recall that the desired fill rate for tankers is 1,000 gpm. Even if the hydrant is capable of flowing 1,000 gpm, there will likely be too much friction loss in the single MDH connected to the hydrant to allow this fill rate to be achievable solely from the residual hydrant pressure. The result will be extremely long fill times, which ultimately decrease the efficiency of the operation. If the organization hooks lines directly to a fire hydrant to refill their mobile water shuttle apparatus, firefighters must remember to connect hydrant gate valves to the hydrant as well. This prevents the hydrant from having to be completely shut down and reopened for each tanker being filled.

The simplest option for increasing the efficiency of a fill operation using medium diameter hose is to add a secondary fill line off the hydrant. The addition of a second line reduces the friction loss in the system and thus increases the fill rate of the tanker (fig. 6–24). Although this may reduce the time it takes to refill the tender, it still may not allow the fill site to reach peak efficiency. This is primarily because many fire departments that rely on hydrants for fill sites also depend on the driver of the shuttle apparatus to make the connections, fill the truck, and then break the connections. More time is spent at the fill site as a result.

Another option is to utilize LDH instead of MDH for the fill operation off the hydrant (fig. 6–25). The increased hose diameter will dramatically reduce the friction loss in the supply line, resulting in higher fill rates off of hydrant pressure. Large diameter lines tend to be more difficult to disconnect after the apparatus is full, however, due to the additional water weight. This typically causes delays in disconnecting from the tender as firefighters fight with the weight on the couplings. In many instances, the reduction in fill time that occurs when LDH is used is lost due to the increased time spent breaking the connections.

Figure 6–24. Using hydrant pressure alone to fill a tender using two medium diameter lines

Figure 6–25. Using hydrant pressure alone to fill a 2,500-gallon tanker with a single LDH line

Some organizations rely on a single hose connection and MDH to refill their tender simply because their apparatus is only equipped with one 2½" direct fill valve. *NFPA 1900* states that mobile water shuttle apparatus must be equipped with an external direct tank fill that is capable of flowing 1,000 gpm from external sources.[24] Although apparatus equipped with a single 2½" direct tank fill may meet the letter of the standard, it is important to remember that there is a tremendous amount of friction loss in MDH when flowing 1,000 gpm. At a flow rate of 1,000 gpm, the friction loss is approximately 200 psi through 100' of 2½" hose and approximately 80 psi through 100' of 3" hose. The hydrant used for the fill operation likely does not have a high enough residual pressure to flow 1,000 gpm and overcome these high friction loss characteristics in a single medium diameter line. This is why when building a tender, it is critical to specify two 2½" direct tank fills on the rear of the apparatus if the organization utilizes MDH for fill operations.

If a fire pump is installed on a tender that only has a single 2½" direct tank fill, the pump panel connections can serve to help adequately fill the apparatus. After connecting the first MDH fill line to the direct tank fill, the second fill line can be connected to the auxiliary intake on the pump panel (fig. 6–26). Prior to charging the fill lines, the auxiliary intake valve as well as the tank fill valve should be opened on the pump panel. At the rear of the apparatus, the direct fill valve is also opened. This will allow the

Figure 6–26. Using the auxiliary intake in conjunction with the direct fill valve to increase the fill rate of this tender

tanker to be refilled at an acceptable rate by splitting the flow between two lines. Depending on how old the tanker is, the total flow rate may not be perfectly split between the direct tank fill and the regular tank fill valves. Remember that on modern apparatus, the tank fill valve must be at least 2" in diameter for apparatus equipped with 1,000-gallon or larger tanks. Therefore, this option will result in suitable fill rates in conjunction with the direct tank fill on the rear of the apparatus. Operators should note that this option is not reserved only for hydrant fill sites. In fact, apparatus equipped with single 2½" direct tank fills should adopt this method as a standard practice.

Pumping the fill site hydrant

For peak efficiency when using a fire hydrant for a fill site, an engine should be used to pump the hydrant. This allows the pumping apparatus to overcome any friction loss in the supply lines used to fill the tankers. In addition, it also provides for more control when opening and closing the discharge valves feeding the tanker. Operators should make sure they perform a heavy hydrant hookup with LDH whenever they are tasked with pumping a hydrant at a fill site (fig. 6–27). This will allow more of the rated capacity of the hydrant to enter the pump due to a reduction in friction loss. These operations will be discussed in more detail in the next chapter.

Figure 6–27. Placing a pumper on the hydrant during fill operations greatly increases the efficiency of the fill site.

If the rural engine company does not carry LDH, it is critical that they hook to the fire hydrant using multiple medium diameter supply lines. This will help ensure that the fill site pumper is capable of filling each tanker at 1,000 gpm. The biggest disadvantage to assigning an engine to the fill site to pump the hydrant is that it ties up one piece of apparatus and potentially its crew. Nonetheless, giving an engine this assignment will drastically increase the efficiency of any rural water supply operation.

Power filling weak hydrants

In a rural environment, the hydrants used for fill site operations tend to be low-flow hydrants due to their location. These hydrants typically deliver well below the 1,000 gpm desired flow rate for use as a fill site. Hydrants in rural communities often provide only 500 gpm flow or even as little as 250 gpm. Filling tankers directly off of these hydrants will result in extremely long fill times. Even if a supply pumper is tasked with connecting to the hydrant and pumping water into the tankers, the desired fill rate will not be met.

In spite of the challenges presented by low flow, however, there is a way that these hydrants can be used efficiently for a fill site operation. Such situations require the pump operator to employ a tactic known as *power filling*. This tactic is similar to using a TurboDraft to fill a dump tank from which the fill site engine drafts. Once a low-flow hydrant is identified, multiple lines should be run from the outlets on the hydrant to fill a nearby dump tank. Each line should be connected to a length of hard sleeve that can then be tucked through the rungs of a ladder that has been leaned against the tank. The hard sleeve acts as a spout that helps direct the water into the tank.[25]

After the tank is full, the fill site engine should establish a prime and prepare to start filling tankers. This technique allows the fill site pumper to draft water from the dump tank and fill each tanker at 1,000 gpm while the hydrant continuously flows into the tank. During a scenario led by the author, this method allowed crews operating from a weak hydrant to fill a 3,000-gallon tanker at a rate of 1,000 gpm from a 250 gpm hydrant. At the end of three minutes, 3,000 gallons had been taken out of the dump tank and pumped into the tanker, and the fire hydrant had replenished the tank with 750 gallons of water.

The key to any successful power filling operation is to maintain a balance between the outgoing water and the incoming water. The hydrant may not be able to refill the tank completely before the next tanker arrives to be refilled. The fill site operator can either reduce the fill rate of the tanker to give the hydrant more time to refill the tank or supplement with the booster tank water.

Sometimes the arrival of the tankers is spread out over longer time intervals. In some situations, the hydrant is able to refill the dump tank, and then water begins overflowing on the ground before the next unit arrives. Water overflowing on the ground in these scenarios is wasted potential. This should signify to the pump operator that an additional dump tank should be deployed and filled off the hydrant. A manifold or multiple gated wyes should be placed on the hydrant outlets to direct water into each tank. In addition, the operator should establish a bridging operation between the two tanks and deploy a jet siphon. Once all of this is complete, the fill site has effectively increased the reserve capacity for filling tankers (fig. 6–28).

Fill Site Layouts

Once the fill site location is identified and selected, the fill site pumper must determine the appropriate fill site layout that optimizes safety and

Figure 6–28. Power filling off of this weak hydrant enabled this fill site pumper to refill each shuttle apparatus at the desired rate of 1,000 gpm.

efficiency. In order to determine the layout, the operator must answer the following questions:

1. Can the mobile water shuttle apparatus safely approach the area near the water source, or do they have to stay out on the main roadway?
2. Will the pump operator be able to see the fill operation from the pump panel?
3. What equipment is available to utilize for the fill operation?
4. How many firefighters are available to assist with the fill operation?

After answering each of these questions, the fill site configuration can be chosen and implemented. Regardless of which fill site layout is utilized, firefighters must remember that fill operations are designed to rapidly refill mobile water shuttle apparatus at a rate of 1,000 gpm. In addition, fill sites should be laid out in an arrangement that allows tankers to easily access the site and leave promptly after being refilled. The firefighters operating the fill site must be proficient with all aspects of a fill operation to ensure peak efficiency is maintained. The two most common fill site configurations are *pumper-controlled fill sites* and *manifold-controlled fill sites.*

Pumper-controlled fill sites

Operators should consider utilizing a pumper-controlled fill site configuration when the following three conditions are met (fig. 6–29):

1. The water source is close enough to the roadway that the fill site pumper can position to access it.
2. The tankers can safely approach the location where the fill site pumper is operating in order to be filled.
3. The operator of the fill site pumper can see the entire fill operation.

This fill site configuration can be utilized for both static and pressurized water sources, but it is the preferred method when using a fire hydrant for a fill operation. After determining that the fill site allows for this layout, the fill site pumper should position in order to safely access the water from the source. Once a water supply is established, hose lines should then be connected to discharges on the fill site pumper. If the department is using MDH, a minimum of two medium diameter lines should be run from discharges on the pumper for filling a single tanker. For departments using LDH for fill operations, it is acceptable to run a single LDH line to each tanker being filled.

Pump operators must remember that during any fill operation, only one tanker should be filled at a time. To maximize fill site efficiency, however, there should be enough hoses on the ground to allow two tankers to be hooked up at any given time. If MDH is being used, this means that four discharges on the pumper should be utilized. For LDH operations, at least

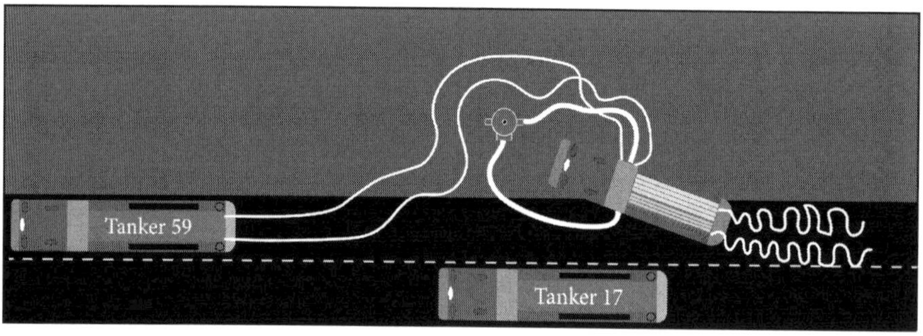

Figure 6–29. A pumper-controlled fill site from a fire hydrant. In this drawing, Tanker 59 is being filled, and Tanker 17 has just arrived at the fill site. The lines off the rear of the engine should be connected to Tanker 17 but should not be charged until after Tanker 59 is completely full.

two discharges will be used, depending on plumbing configurations. Although the fill site pumper should only fill one tanker at a time, having additional hoses on the ground allows fill site personnel to make connections to the second tanker while the first one is being filled. Once the first tanker is completely filled, the operator can simply open the appropriate discharges to begin filling the second tanker.

The major advantage of a pumper-controlled fill site is that it allows the pump operator to manage the opening and closing of the discharge valves filling each tender. This not only reduces the number of personnel needed to operate the fill site, but it also allows for the discharge valves to be opened and closed in a more controlled manner. In order for this type of fill operation to be successful, pump operators must be able to see the entire fill operation. This allows them to know when they can safely open the discharges feeding the tenders and also when to shut down these discharges after the tender is full. If there is a distance between where the fill site pumper is located and where the tenders can safely access the fill site, a manifold-controlled fill site operation should be employed instead.

Manifold-controlled fill sites

When dealing with static water sources in the rural environment, it is not uncommon for the water source to be located a long distance from the main roadway. In addition, the path leading to the source may be narrow, unpaved, and treacherous for heavy apparatus. When faced with these scenarios, the operator should elect to perform a manifold-controlled fill site operation. This will enable the fill site pumper to access the water source while keeping the tenders out on the main roadway (fig. 6–30).

This operation requires the fill site pumper to drop a supply line at the intersection of the main roadway and the path leading to the water source. A manifold or large diameter appliance with multiple discharges should be left at the intersection and connected to the end of the supply line (fig. 6–31). Ideally, the fill site engine lays LDH to the source. If the pumper is only equipped with MDH, however, it is highly recommended that dual supply lines be laid to ensure the desired fill rate is met. This will require both supply lines to enter into the manifold via a siamese. Prior to laying out toward the water source, the fill site pumper should leave lengths of hose to be connected on the outlet side of the manifold for the fill operation. Arriving at the water source, the fill site pumper should establish a draft and connect the supply line to a discharge. When the personnel at the

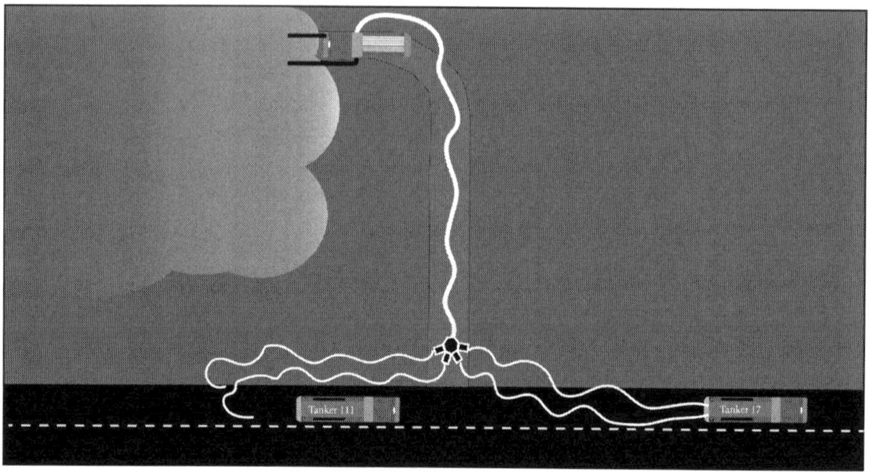

Figure 6–30. A manifold-controlled fill site operation. Note that even though two tankers can be hooked up in this configuration, only one tanker is filled at a time.

Figure 6–31. A manifold-controlled fill operation

manifold confirm that they are ready for water, the fill site pumper can charge the supply line leading up to the manifold.

After charging the supply line, the operator of the fill site pumper should calculate the required pump discharge pressure for the hose layout and set the pump to maintain that pressure. To determine the pump pressure, the operator should calculate the friction loss for 1,000 gpm in the supply line laid. As a general guideline, at least 10 psi should be added for the appliance loss in the manifold. The exact appliance loss value for the manifold will vary depending on the particular appliance utilized. The operator should then calculate the friction loss in each line coming off the manifold and feeding the tanker. If dual MDH lines are being utilized, the operator must remember to split the total flow rate between both supply lines feeding the tanker, i.e., 500 gpm in each line. Finally, the operator should add 20 psi to account for any plumbing loss within the direct fill valves of the tanker and to provide some residual pressure entering the tank.

Firefighters operating at the fill site should think of the manifold as an extension of the pump panel. The manifold is used to control the flow of water to the tankers being filled. The advantage of this is that the pump operator does not need to worry about when to open or close discharge valves. The pump operator's primary concerns will be to maintain the appropriate discharge pressure throughout the entire operation and to ensure that the prime is not lost between tanker fills. One of the disadvantages of this configuration is that it requires one additional person at the fill site in order to operate the manifold.

When a manifold-controlled fill site operation is employed, the operator must be concerned with overpressurization of the supply line feeding the manifold. This usually occurs once the tanker is full, and the flow goes from 1,000 gpm to 0 gpm as the valves of the manifold are shut. If the appropriate measures are not taken, overpressurization could result in an injury due to a burst supply line. For this reason, it is critically important for fire departments to ensure they are using manifolds that are equipped with a built-in pressure relief valve (fig. 6–32). Firefighters operating the manifold should pay close attention to where the blow off for the relief valve is pointed. In addition, it is critically important for the firefighters operating the manifold to ensure that they shut the valves of the manifold slowly. The pump operator can also help prevent excessive pressures from building in the supply line by ensuring that their pressure relief devices at the pump are appropriately set.

Determining the Needed Number of Tenders

In order to provide the needed fire flow at any rural fire scene, the incident commander must ensure that there are enough tenders involved in the shuttle route. Many organizations achieve this by dispatching a specific

Figure 6–32. Be mindful of the manifold relief valve opening once the fill lines are shut down.

number of tenders on each rural water assignment. This ensures that there is a sufficient number of apparatus responding to manage the water supply needs for most rural fires. Nonetheless, it is critically important for the water supply officer to be able to calculate whether there is a sufficient number of apparatus available to meet the fire flow needs of an escalating incident.

There are several ways to evaluate the performance of mobile water supply apparatus. Some departments will actually perform physical tests to determine the efficiency of their mobile water supply apparatus. While this offers detailed insight into the capabilities of a particular organization's shuttle apparatus, it may not be feasible for a department to perform this type of testing. Instead, some fire departments utilize calculations to estimate the capabilities of their apparatus. The first calculation used determines the *travel time* from the dump site to the fill site and back:[26]

$$\textit{Travel Time} = 0.65 + (1.7)(D)$$

where

0.65 = Constant used for acceleration and deceleration of the vehicle
1.7 = Constant for an average speed of 35 mph
D = The total round-trip distance in miles

It is important to note that using this equation assumes the mobile water shuttle apparatus is able to maintain an average speed of 35 mph. Road conditions, legal speed limits, and the vehicle's capabilities will all dictate whether the apparatus can travel at a speed above or below 35 mph. Correction factors can be used to determine the travel time if the vehicle speed must be adjusted. The reader should note that all charts and tables in this text assume an average vehicle speed of 35 mph.

The next component that should be calculated is the *handling time*. The handling time is comprised of the time the tanker spends at the dump site off-loading and the time it spends at the fill site refilling. Firefighters must remember this is the only variable that they can directly impact. By employing efficient dump and fill site tactics, the handling times can be dramatically reduced, and the efficiency of the operation can be increased. Handling time in minutes can be calculated by the following equations for fill site time and dump site time, which are also given in minutes:

$$Handling\ Time = Fill\ Site\ Time + Dump\ Site\ Time$$

$$Fill\ Site\ Time = PT + \frac{TC}{FR}$$

$$Dump\ Site\ Time = PT + \frac{TC}{DR}$$

where

PT = Positioning time in minutes
TC = Tender capacity in gallons
FR = Fill rate in gallons per minute
DR = Dump rate in gallons per minute

The positioning time at the fill site not only includes the time needed to position the tanker, but also the time it takes to make and break the connections to the truck. The goal is for this to be completed within 30 seconds, and it should not take more than one minute. The positioning time at the dump site reflects the time it takes for the apparatus to appropriately position near the dump tank prior to off-loading. If the dump site is laid out efficiently, this time should be minimal.

The final component to determining the needed number of tankers for an operation is to calculate the continuous tanker flow rate for each apparatus. This value includes all the variables previously calculated and gives the continuous flow rate in gallons per minute each tender is providing for use at the scene. The water supply officer can then use this number to determine how many tankers are needed to provide a specific volume of water for the operation. The calculations for determining the *continuous tanker flow rate* in gallons per minute and *needed number of tankers* are as follows:

$$\textit{Continuous Tanker Flow Rate} = \frac{0.9 \times \textit{Tank Capacity}}{\textit{Travel Time} + \textit{Handling Time}}$$

where

Tank capacity is given in gallons,
Travel time is given in minutes, and
Handling time is given in minutes.

$$\textit{Needed Number of Tankers} = \frac{\textit{Needed Fire Flow}}{\textit{Continuous Tanker Flow Rate}}$$

where

Needed fire flow is given in gallons per minute, and
Continuous tanker flow rate is given in gallons per minute.

While having an understanding of these calculations is important, it may not be possible to perform them on the fire scene. In order to simplify these calculations, the fire department should utilize a continuous tanker flow chart similar to table 6–1. These values should give the reader an idea of the information that these tools can provide. The fire department should take this information and adjust it to fit their specific requirements, such as the tender sizes used in their organization and any necessary adjustments to average vehicle speed.

Large-Scale Fill Site Operations

Certain fire scenes may require more water than a simple fill operation is capable of delivering. One option for dealing with this situation is to

Table 6–1. Continuous tanker flow rates based on various fill rates and round-trip travel distances

Tanker Capacity (gal)	Tanker Fill Rate (gpm)	1 Mile CTFR (gpm)	2 Mile CTFR (gpm)	3 Mile CTFR (gpm)	4 Mile CTFR (gpm)	5 Mile CTFR (gpm)	10 Mile CTFR (gpm)	15 Mile CTFR (gpm)	20 Mile CTFR (gpm)
1,500	1,000	213	168	138	118	103	62	45	35
	750	197	158	132	113	99	61	44	34
	500	172	141	120	104	92	58	43	34
2,000	1,000	245	199	167	145	127	79	58	45
	750	225	185	158	137	121	77	57	45
	500	193	163	141	125	111	73	54	43
2,500	1,000	269	224	191	167	149	95	70	55
	750	245	207	179	158	141	92	68	54
	500	207	179	158	141	127	86	65	52
3,000	1,000	289	244	212	187	167	110	81	65
	750	261	224	196	175	157	105	79	63
	500	219	192	171	155	141	98	75	60
3,500	1,000	304	261	229	204	184	123	92	74
	750	274	238	211	190	172	117	89	72
	500	227	203	183	166	153	108	84	68
4,000	1,000	317	276	244	219	198	135	102	82
	750	284	250	224	202	185	129	99	80
	500	235	211	192	176	163	117	92	76
4,500	1,000	328	288	257	232	211	146	112	91
	750	292	260	235	214	196	139	108	88
	500	240	218	200	185	171	126	100	82
5,000	1,000	337	299	269	244	223	157	121	99
	750	300	269	244	224	206	148	116	95
	500	245	224	207	192	179	134	107	89
5,500	1,000	345	308	279	254	234	167	130	106
	750	306	277	253	233	215	157	124	102
	500	249	230	213	198	186	141	113	95

Note: CTFR is the continuous tanker fill rate.

establish additional fill sites at other nearby sources. If the original source is bountiful and close enough to the fire scene, multiple fill site pumpers may be assigned to the fill operation at the original source. This section will focus on options for implementing large-scale, long-duration fill operations. The water supply officer and incident commander must ensure that sufficient resources are committed to these fill operations in order for them to be successful. Equally important is ensuring that the appropriate number of tankers will be available to provide the needed fire flow.

Expected fill capacity of a single fill site pumper

Recognizing the realistic fill rate of a single pumper is the first step in determining how many additional fill site engines will be needed for more complex operations. The two main pillars of an efficient fill operation are to fill each tanker at a rate of 1,000 gpm and to only fill one tanker at a time. Attempting to fill two tankers at once would require the fill site pumper to flow a total of 2,000 gpm, but realistically, the fill rate would likely be only around 500 gpm for each tanker. This ultimately keeps both tankers at the fill site for a much longer period of time.

The water supply officer and the operator of the fill site pumper should expect a realistic flow rate from a single fill site pumper of 1,000 gpm, regardless of the rated capacity of the pump. This general guideline is based on the specific conditions under which fire pumps are tested. For example, a 2,000 gpm pumper must be capable of flowing this volume of water from a draft, at a pump discharge pressure of 150 psi, with no more than two intakes, each equipped with 20' of 6" hard sleeve, and no more than 6' of lift. But in a rural setting, many of these pump testing criteria may not be met. For instance, any of the following could be encountered during operations in a rural setting:

- The lift present may be more than what is accepted for a pump test.
- Reaching the water source may require a long reach operation that uses more than 20' of hard sleeve.
- Initially, there may only be enough hard sleeve present to utilize a single intake.
- Depending on the fill site layout and hose lines utilized, the pump discharge pressures may exceed 150 psi.
- The apparatus may run out of horsepower before delivering the 2,000 gpm necessary to efficiently fill two tenders at once.

For these reasons, a single fill site pumper should be expected to reliably deliver 1,000 gpm for a fill operation (fig. 6–33). This allows the fill site pumper to refill tankers at the appropriate rate under more challenging actual conditions. If the water supply officer determines that the fill site needs to be capable of filling multiple tankers at a time, an additional fill site pumper must be incorporated. In other words, a fill site capable of 2,000 gpm will require two fill site pumpers. A good indication of the need for an additional fill site pumper is if the tankers are stacking up at the fill site waiting to be filled. The water supply officer must anticipate the need for additional fill site pumpers early due to the time it will take for these units to mobilize and deploy.

Once it is determined that the needed fire flow at the scene is outpacing the ability of the fill site, the proper tactic for increasing the fill capacity of the operation must be chosen. Generally speaking, the water supply officer has two options available. The first option is to open an additional fill site from a different source. If the original source is bountiful, however, multiple fill site pumpers can be assigned to operate from the original source.

Opening additional fill sites

Opening an additional fill site requires that a suitable water source be located in the vicinity of the incident scene. Opening a secondary or tertiary fill site that is a significant distance away from the dump site will result in

Figure 6–33. Each fill site pumper should be considered good for up to 1,000 gpm.

reduced efficiency of the shuttle operation due to the increased travel distance. In order to appropriately gauge this, the water supply should reference a tanker capacity chart similar to the one outlined in table 6–1. It is also important to identify whether the additional source is a fire hydrant or a static water source. This will determine the equipment needed at the source and may help to determine the layout of the fill site. Static sources should be evaluated to determine if they contain a sufficient amount of water to support the operation. Hydrant sources must be capable of providing at least 1,000 gpm water flow.

The option of opening an additional fill site is not reserved only for large-scale fires in the rural environment. This tactic should also be considered if the original fill site is not performing efficiently or if tankers are not being filled at the desired fill rate. Opening an additional fill site in these situations will help increase the efficiency of the operation. If the original fill site is failing to fill tankers at 1,000 gpm, it does not necessarily mean that this fill site should be closed or abandoned, especially if this source is close to the scene. The original low-flow fill site can be used in conjunction with the secondary fill site to help distribute the workload of the entire operation.

The water supply officer must consider implementing multiple fill sites when sources with limited water are used initially. This includes scenarios where cisterns, swimming pools, or other small static sources were initially used. Although these sources may have been the closest water to the fire scene, they may be depleted quickly on larger fires. Using these sources for the initial operation may be warranted, but the water supply officer must identify an alternative source early in the incident in case it is needed.

Once an additional fill site is put into service, the water supply officer must be careful to assign each tender a particular fill site for refilling. This will prevent confusion and will keep one of the fill sites from becoming overwhelmed with a backlog of tenders. When determining which site each tanker will be assigned to, the water supply officer must consider such factors as the capacity of each tender, the distance each fill site is from the dump site, the fill rate of each fill site, and how easily the tenders will be able to access each site.

Multiple fill site pumpers at the original source

Deploying multiple fill site engines to a single water source is a fantastic option when the original source is close to the fire scene and has an abundance of water. This will not only allow more of the water in the source to be utilized, but it will also minimize the travel distance of the shuttle route. By simply adding one additional fill site pumper to the original source, the efficiency of the filling operation can potentially double. Prior to employing this tactic, the fill site officer must first determine if there is enough water in the source to support multiple pumpers and then decide how additional pumpers will access the water (fig. 6–34).

Typically, the biggest challenge when dealing with bountiful sources tends to be access for the additional fill site pumpers. This is usually because the initial fill site pumper has already positioned at the most advantageous location. In these instances, the operators of the subsequent fill site pumpers should be prepared to perform either a long reach or TurboDraft operation to access the water (fig. 6–35). Later-arriving fill site pumpers that elect to perform a TurboDraft operation should strongly consider using the device to fill a dump tank. This allows the additional fill site pumper to position where it is best suited, while still having access to the water (fig. 6–36).

The layout of a multiple pumper fill site will be dependent on the same variables previously discussed, but typically it is dictated by how each pumper is able to access the water. Some scenarios may allow both fill site engines to perform a pumper-controlled fill operation. Other scenarios may allow the initial fill site pumper to employ a pumper-controlled fill site

Figure 6–34. Two pumpers are operating from a bountiful water source, each utilizing a twin tube drafting configuration.

Figure 6–35. These three pumpers have established a 3,000 gpm capable fill site. The first two pumpers down the ramp were able to use conventional drafting techniques. The third engine performed a tandem TurboDraft operation.

Figure 6–36. This pumper was the third engine to arrive at the fill site. In order to access the water, two TurboDrafts were used to fill this dump tank, allowing the engine to draft from it.

layout while the second engine performs a manifold-controlled fill site operation. Still other water sources may dictate that each fill site pumper lay their own supply line and perform a manifold-controlled fill site operation. As tankers arrive at a fill site that has multiple fill site pumpers in operation, they should be directed to a particular station for refilling. In addition, it is critical to ensure that these operations are appropriately staffed in order to manage each filling station at the fill site.

Filling large capacity tankers

Some jurisdictions in the United States utilize mobile water shuttle apparatus with tank capacities of 4,000 gallons or greater. When these large capacity tankers are utilized in a shuttle operation, they can potentially affect the efficiency of the operation in a negative way. This is especially true when the other tankers in the shuttle operation are smaller in capacity. Smaller tenders will typically stack up waiting for their turn while the large capacity tender is refilled (fig. 6–37).

If larger capacity tankers are required to participate in the shuttle route, there are a couple of options available to maintain the efficiency of the operation. One option to consider is to open a separate fill site for the large capacity tankers, which will help prevent bottlenecks at the primary fill site. A potential downfall of this option is that the secondary fill site may be located at a greater distance from the fire scene. In these instances, the water supply officer must decide where to send each tanker to be refilled. While there are several variables that should be considered, emphasis must be placed on determining what the continuous flow rate of each tanker will be if the tanker is sent to a particular fill site.

Figure 6–37. Large capacity tenders will typically cause a bottleneck when they are refilled at the same site as smaller tenders.

For example, consider a scenario in which two 2,000-gallon tankers and one 4,000-gallon tanker are operating in a shuttle route. The primary fill site is a 2-mile round trip from the dump site and is capable of filling at 1,000 gpm. The closest secondary fill site has a round-trip distance of 10 miles and is also capable of filling apparatus at 1,000 gpm. In this scenario, the water supply officer must determine where to send the large tanker and where to send the smaller tankers. Referring to table 6–1, the operator can determine what the continuous tanker flow rate (CTFR) of each rig will be and where to send each apparatus.

Option 1: 1,000 gpm fill rate
2,000 gal tanker CTFR at 2 miles = 199 gpm × 2 = 398 gpm
4,000 gal tanker CTFR at 10 miles = 135 gpm
Total CTFR available at the scene = 533 gpm

Option 2: 1,000 gpm fill rate
2,000 gal tanker CTFR at 10 miles = 79 gpm × 2 = 158 gpm
4,000 gal tanker CTFR at 2 miles = 276 gpm
Total CTFR available at the scene = 434 gpm

In this scenario, the best option would be to send the large capacity tanker to be filled at the further fill site and keep both of the smaller tankers at the primary fill site. Configuring the shuttle operation in this fashion allows the flow rate at the fire scene to be increased by nearly 100 gpm. Situations may occur, however, in which the closer fill site can only provide a fill rate of 500 gpm, while the further fill site can maintain a 1,000 gpm flow. In this scenario, it is critically important for the water supply officer to determine which fill site assignments for the tenders offer the maximum efficiency.

Option 1: 1,000 gpm and 500 gpm fill rates
2,000 gal tanker CTFR filling at 500 gpm, 2 miles = 163 gpm × 2 = 326 gpm
4,000 gal tanker CTFR filling at 1,000 gpm, 10 miles = 135 gpm
Total CTFR available at the scene = 461 gpm

Option 2: 1,000 gpm and 500 gpm fill rates
2,000 gal tanker CTFR filling at 1,000 gpm, 10 miles = 79 gpm × 2 = 158 gpm
4,000 gal tanker CTFR filling at 500 gpm, 2 miles = 211 gpm

Total CTFR available at the scene = 369 gpm

In this scenario, we see that it is advantageous to send the larger capacity tanker to the fill site located further away and keep the smaller apparatus closer to the scene. It is important to note that these examples assume that there are no other potential hazards present that would make sending the larger apparatus a further distance impractical. Generally speaking, when multiple fill sites are in operation, it is most beneficial to send large capacity tankers to the distant fill sites while smaller tankers refill closer to the scene.[27]

It may not be realistic to open an additional fill site just to support the filling of a large capacity tanker. If the primary source is bountiful and can support multiple pumpers, a simple solution could be to assign an additional fill site pumper to the fill site. The second fill site pumper can be used solely to refill the large capacity tanker without impeding the refilling ability of the smaller tankers (fig. 6–38).

If the larger tanker is equipped with a fire pump, it is possible to use two fill site engines to fill the apparatus simultaneously. This would allow the fill time to essentially be cut in half since each fill site pumper is flowing 1,000 gpm. This tactic can be accomplished by either a pumper-controlled

Figure 6–38. If the source is plentiful, one fill site engine can be assigned to refill small capacity tankers while a second pumper fills the larger capacity apparatus.

or manifold-controlled fill site operation. Ideally, MDH lines should be connected to the large capacity tender for this operation. This means that four lines will ultimately be connected to the truck being filled (fig. 6–39).

The first step in this operation is to connect the fill lines from the first fill site engine to the direct fill valves on the rear of the tender. In this configuration, the first engine fills the apparatus as usual. The fill lines from the second pumper are then brought to the pump panel of the tender to be connected. One line should be connected to the auxiliary intake on the apparatus, while the other should be connected to a 2½" discharge. Prior to charging the fill lines, several valves must be opened: (1) the auxiliary intake valve, (2) the valve of the discharge with the corresponding fill line attached, and (3) the tank fill valve. When the fill lines from the second pumper are charged, the water will flow through the plumbing of the pump and through the tank fill valve to help refill the apparatus. Since each fill site pumper is striving to fill at a rate of 1,000 gpm, this configuration should enable the tanker to be refilled at a rate of 2,000 gpm. When used properly, this method for refilling tankers can rapidly decrease the fill times for larger capacity tankers. The beauty of this operation is that it requires no change in the pump pressures of the fill site engines operating from the source. In addition, after the larger capacity truck is full, two smaller capacity tankers can be filled immediately afterward.

Figure 6–39. This 4,000-gallon tanker was filled in two minutes using two fill site pumpers.

Firefighters must be cautious when employing this tactic and only utilize it if they know their apparatus is capable of being refilled in this fashion. At no point should the flow rate through any individual fill valve exceed 1,000 gpm into the tank of the apparatus. In addition, it is generally recommended that the pump pressure at 1,000 gpm entering the tank not exceed 100 psi. Failure to follow these guidelines can cause damage to the apparatus and injury or death to fill site personnel. This tactic is not recommended to be utilized on retrofit or homebuilt mobile water shuttle apparatus.

Study Questions

1. Which of the following should be considered when preplanning rural water fill sites?

    a. Accessibility of the source
    b. Personnel available
    c. Tanker capacity
    d. Ability of the pump operator

2. When using a dry hydrant, the operator must ensure which of the following?

    a. The hydrant is completely turned on.
    b. Only a 10' section of hard sleeve is used.
    c. The dry hydrant is back flushed prior to attempting a prime.
    d. The gaskets are removed from the hard sleeve.

3. Which of the following is one of the easiest ways to get a TurboDraft into the water source?

    a. Tying it to a ladder and sliding it into the source
    b. Attaching a float and throwing/pushing it into the water
    c. Getting into the water and placing it
    d. Using an aerial ladder to lower it into the water

4. It is generally recommended that TurboDrafts be pumped at which of the following pump discharge pressures?

    a. 100 psi
    b. 150 psi

c. 200 psi
d. 300 psi

5. Which of the following is an option for increasing the flow capacity of a TurboDraft operation?

   a. Decrease the hose diameter.
   b. Use two pumpers to power a single TurboDraft.
   c. Use a TurboDraft to fill a dump tank, then draft from it.
   d. All of the above are options.

6. Which of the following is the desired fill rate for mobile water supply apparatus?

   a. 500 gpm
   b. 750 gpm
   c. 1,000 gpm
   d. 2,000 gpm

7. How should the fill site be configured when the operator of the fill site engine can see the filling operation?

   a. As a pumper-controlled fill site.
   b. As a manifold-controlled fill site.
   c. As a siamese-controlled fill site.
   d. No fill site pumper is needed.

8. Realistically, what fill rate is a single fill site engine capable of?

   a. 500 gpm
   b. 1,000 gpm
   c. 2,000 gpm
   d. Whatever its rated pump capacity is

9. When opening an additional fill site, which of the following statements is true?

   a. It should be twice as far from the scene as the original.
   b. It must be capable of providing 2,000 gpm.
   c. It should be the next closest source to the scene, aside from the primary fill site.
   d. It must be a fire hydrant.

10. Generally speaking, smaller tankers should be filled from sources _______________, while large capacity tankers should refill at sources __________________.

    a. Far from the scene; closer to the scene
    b. Close to the scene; further from the scene
    c. That are static; that are fire hydrants
    d. That are fire hydrants; that are static

7

HYDRANT OPERATIONS

Fire hydrants serve as the primary water source for many fire departments across the United States. Hydrants are usually capable of providing more than enough water for residential firefighting operations. The advent of large diameter hose has caused many organizations to simply hook their supply line directly to the fire hydrant and use hydrant pressure alone to supply water to these fire scenes. While this may be sufficient for most residential fires, operating in this fashion will severely limit the flow capability of the scene if the incident expands beyond a bread-and-butter operation. In addition, using hydrant pressure alone to supply a fire scene is typically an insufficient means of providing the needed fire flow for most commercial fires.

As discussed in chapter 3, the best option for maximizing the capability of a fire hydrant is to assign an engine to pump the hydrant. This supply pumper should perform a *heavy hydrant hookup* in order to increase the total area feeding the pump and thus reduce the friction loss on the intake side of the system. In this chapter, we will discuss why this process has such a profound impact on the availability of water from a municipal water supply system. Options for overcoming the most common hydrant challenges will also be introduced. These concepts will become critical components during the more complex water supply operations discussed in the remaining chapters.

Rated Flow Capacities of Fire Hydrants

Maximizing the water available from a fire hydrant requires the pump operator to recognize how a hydrant is designed to discharge its rated capacity. Fire hydrants flow their rated capacity under two conditions. The first condition is that the residual main pressure is 20 psi. The second condition is that all the outlets are open. This section will discuss why it is critically important for the pump operator to ensure these conditions are met when attempting to maximize the water from a hydrant. We will also examine how maximum flow rates are calculated through individual fire hydrants as well as water mains.

The 20 psi rule

Pump operators have likely been told at some point in their careers not to let the master intake gauge drop below 20 psi when operating from a fire hydrant. There is often a lot of controversy as to why this rule exists and what impact it has on the pump operator. Both the American Water Works Association (AWWA) and *NFPA 291: Recommended Practice for Water Flow Testing and Marking of Hydrants* specifically state that at no point should the pressure within a municipal water supply system drop below 20 psi.[1] Dropping the residual pressure in the system below this point could potentially cause damage to components of the municipal water system. The more concerning issue is the potential for polluted water to be siphoned back into the distribution system from other interconnected sources if the system pressure drops below 20 psi.[2] The entire distribution system could be contaminated with nonpotable water, creating a public health problem.

As we can see, failure to maintain an appropriate residual pressure in the water distribution system can have a catastrophic impact. Historically, firefighters have relied on their master intake gauge to determine whether or not they were reaching this danger point. Relying solely on the master intake gauge may not give the pump operator full insight into the situation, however. In order to fully understand the significance of the 20 psi rule, the pump operator must first distinguish between residual hydrant pressure and residual intake pressure.

Residual hydrant pressure is the remaining pressure at the hydrant when water is flowing out of each hydrant outlet. The residual hydrant pressure should be considered the closest representation of the pressure within the water main itself since there is no other feasible way for the pump operator to determine the actual residual pressure in the water main. As previously mentioned, residual intake pressure refers to the remaining pressure at the eye of the pump's impeller when water is being discharged from the pump.[3] Residual intake pressure is determined by looking at the master intake/compound gauge on the pumper.

The laws of hydraulics state that friction loss is present within a system any time water travels through a conduit between two points. In addition, when the flow rate through the system increases, the friction loss in that system increases at an exponential rate. Because of these hydraulics laws, we know that when water flows out of a fire hydrant, through supply hose, and into a fire department pumper, there is friction loss between these two points. Sources of the friction loss in the system include the hydrant, the appliances on the hydrant, the supply hose, and the intake valve/plumbing of the apparatus. As the discharge rate from the supply pumper increases, so does the friction loss between the hydrant and the pumper's intake.

Traditionally, the master intake gauge has been the primary tool available to the pump operator to measure pressure on the intake side of the pump. Since it is critically important to maintain a residual pressure in the water distribution system of 20 psi, most operators consider the hydrant maxed out once the master intake gauge shows a residual intake pressure of 20 psi. The operator must recognize that a residual intake pressure reading of 20 psi does not necessarily represent a residual hydrant pressure of 20 psi. This is due to the friction loss between the hydrant and the pumper.

Determining the residual hydrant pressure requires a gauge to be attached to the hydrant initially. Certain manufacturers produce a 2½" hydrant gate valve that has a built-in pressure gauge on the intake side of the valve.[4] This allows the pump operator to note the residual hydrant pressure regardless of whether the gate valve is open or shut (fig. 7–1). When flowing large volumes of water from a fire hydrant, the pump operator must pay attention to both the residual intake pressure and the residual hydrant pressure. At higher flow rates, the residual intake pressure will likely be lower than the residual hydrant pressure since the master intake gauge reads pressure downstream in the system. Because of this, it is possible for the residual intake pressure on the apparatus to show 20 psi while the residual hydrant pressure reading is much higher. The difference in these two readings represents the friction loss present between the hydrant and the eye of the impeller. Therefore, it is completely acceptable to drop the residual intake pressure at the pumper below 20 psi as long as the residual hydrant pressure remains at 20 psi or higher. This will enable the supply pumper to achieve more of the rated capacity of the hydrant.

Figure 7–1. A hydrant valve with a built-in pressure gauge that informs the operator of the residual hydrant pressure

The use of multiple hydrant outlets

In order to fully maximize the volumetric flow rate from a fire hydrant, the pump operator must be prepared to utilize multiple hydrant outlets. The most common type of *dry-barrel hydrant* in the United States is equipped with three outlets: one 4½" steamer port and two 2½" side ports. Each dry-barrel fire hydrant is equipped with an internal main valve that either opens or closes the device. These internal main valves are typically either 4½" or 5¼" in diameter.[5]

The water distribution system of a municipality is evaluated as a whole during a Public Protection Classification Review conducted by the Insurance Services Office (ISO). The number, configuration, and flow capacities of the fire hydrants within the municipality have a significant impact on the classification results. ISO will give maximum credit for each hydrant that is equipped with a 4½" steamer port, two 2½" side ports, and a 5¼" internal main valve.[6] Hydrants with fewer outlets or smaller main valves will receive prorated credit. Each fire hydrant is connected to the mains, which feed them using a piece of pipe known as a *branch* or *lateral*. As an industry best practice, the AWWA recommends that the branch/lateral connecting the hydrant to the water main be at least 6" in diameter.[7]

The pump operator can use this information regarding hydrant components to identify the limiting factor that will affect their ability to maximize the hydrant. In order to do this, we must first convert the diameter of each hydrant component into its respective area value. These converted values are listed in tables 7–1 and 7–2. A review of table 7–2 indicates that when the 4½" steamer port and one 2½" side port are utilized together, their combined total area opening is the equivalent to the area of the 5¼" main valve. This makes the outlet area equal to the smallest internal component of the hydrant itself. The pump operator should consider using at least two outlets, or *double tapping* the hydrant, as the minimum standard moving forward for maximizing hydrant flows.[8]

Table 7–1. Hydrant component dimensions

Hydrant component	Diameter (in.)	Area (in.^{2})
Steamer port	4.576	16.446
Side port	2.576	5.212
Hydrant main valve	5.250	21.648
Branch/Lateral	6.000	28.274

Note: Hydrant outlet diameters are measured to the nearest thousandth of an inch.
Source: A. Soccodato, "Heavy Hydrant Hookups: Maximizing Hydrant Flows, Part 1," *Fire Apparatus & Emergency Equipment* 27, no. 7 (2022).

Table 7–2. Comparison of total area openings

Hydrant component/Configuration	Total area (in.^{2})
Steamer port and one side port	21.658
5¼" hydrant main valve	21.648
Steamer port and both side ports	26.870
Branch/Lateral	28.274

Source: A. Soccodato, "Heavy Hydrant Hookups: Maximizing Hydrant Flows, Part 1," *Fire Apparatus & Emergency Equipment* 27, no. 7 (2022).

When the operator elects to utilize the steamer port and both 2½" side ports, the total outlet area exceeds the area of the hydrant main valve. Even so, the 6" branch/lateral still has the largest area opening. This makes the main valve of the hydrant the limiting factor with regard to achieving the maximum flow rate through the hydrant, not the outlet configuration established by the pump operator.[9] During heavy flow fires of long duration, the operator should be prepared to maximize the hydrant using all the outlets provided. While the number of hydrant outlets utilized has a significant impact on the water availability from a hydrant, the operator must recognize that it is only part of the equation. Equally important to achieving these larger flows is the diameter of the supply line used to connect the pumper to the hydrant. This will be discussed later in this chapter.

Calculating flow rates through individual hydrants and water mains

There are two accepted methods for the flow testing of fire hydrants, but it must be understood that they provide slightly different data. The first method is used to determine the specific flow rate from an individual fire hydrant. This procedure requires a *cap gauge*, or a pressure gauge installed on a blind cap for hydrant testing purposes, to be placed on one of the 2½" outlets. The hydrant is then opened, and the static pressure is measured. After recording the static pressure, the hydrant is shut, and the other 2½" outlet is removed. The hydrant is then reopened, and a pitot tube is used to determine the flow rate exiting the outlet of the hydrant. After the pitot reading is captured, the residual pressure reading on the cap gauge is recorded. The information gathered is then entered into an equation, known as the *Hazen-Williams formula*, to determine what the flow rate through this individual hydrant will be at a residual pressure of 20 psi. While this information is useful, it does not provide the fire department with a full picture regarding the capacity of the municipal water supply system feeding the hydrant tested.

A slightly different approach must be taken in order to determine the main capacity of a specific portion of the municipal water supply system. This method requires the use of at least two fire hydrants that are located on the same main. One of these hydrants will be identified as the *test hydrant*, while the other will be the *flow hydrant*. At the test hydrant, a cap gauge is connected to one of the 2½" outlets, the hydrant is opened, and the static pressure is recorded. At the flow hydrant, one of the 2½" outlet caps should be removed, and the hydrant should be opened. Once water is flowing out of the flow hydrant, a pitot tube should be inserted into the stream and the value recorded. Simultaneously, the residual pressure should be recorded at the test hydrant.

A general guideline for this process is to flow enough water for the residual pressure to drop at least 10 psi below the static pressure originally recorded.[10] This will provide an accurate depiction of the flow capacity of the main being tested. If a single outlet on the flow hydrant does not sufficiently drop the residual pressure, opening the second 2½" outlet or an additional flow hydrant must be considered. It is also important to note that prior to the test, firefighters should inspect the outlets of the flow hydrant to determine if a coefficient must be used when calculating the flow results. For a more in-depth explanation of hydrant flow testing, the reader should refer to the AWWA's *Fire Hydrants: Installation, Maintenance, and Field-Testing—M17* manual and *NFPA 291*.[11]

After performing the flow test and gathering the necessary data, the capacity of the main can be calculated using the Hazen-Williams formula. This formula determines what the expected flow rate will be in gallons per minute at a residual pressure of 20 psi:[12]

$$Q_R = Q_F \times \frac{h_r^{0.54}}{h_f^{0.54}}$$

where

Q_R = The flow rate available at the desired residual pressure (20 psi in most cases), gallons per minute

Q_F = The flow rate during the test, gallons per minute

h_r = The difference between the static pressure recorded and the desired residual pressure being calculated

h_f = The difference between the static pressure recorded and the residual pressure recorded during the flow testing

Example 7–1. You are performing hydrant testing to determine the capacity of a particular main in your response area. Calculate the capacity of the main based on the data listed below:

Static pressure: 70 psi

Residual pressure: 58 psi

Flow rate during test: 1,200 gpm

$$Q_R = Q_F \times \frac{h_r^{0.54}}{h_f^{0.54}}$$

$$Q_R = 1{,}200 \times \frac{(70-20)^{0.54}}{(70-58)^{0.54}}$$

$$Q_R = 1{,}200 \times \frac{50^{0.54}}{12^{0.54}}$$

$$Q_R = 1{,}200 \times \frac{8.27}{3.83}$$

$$Q_R = 2{,}591 \text{ gpm}$$

In this example, the water main feeding the hydrants used for the testing is capable of supplying 2,591 gpm. It is critically important to understand that this flow rate may be more than a single fire hydrant is capable of supplying. Achieving this flow rate on an actual incident could require multiple supply engines to be connected to multiple hydrants on this particular main.

Firefighters should have a solid understanding of which method their jurisdiction utilizes to determine hydrant flow rates. If single hydrant flow testing is performed, the calculated values may not be representative of what the main is actually capable of providing.[13] Conversely, main capacity flow tests may yield flow rates that are not realistic utilizing a single fire hydrant. Understanding why there is a variance between these two methods can help the operator to determine the options for maximizing their water supply.

Based on the results from the hydrant flow tests performed, some municipalities elect to paint their fire hydrants a certain color based on their flow capacity. *NFPA 291* contains a recommended color-coding system to identify hydrants based on their flow rate (see table 7–3).[14] Pump operators should not base their available flow rate from a hydrant solely on the hydrant color. Several factors exist that can influence the actual flow rate available from a particular fire hydrant. Firefighters must remember that water availability in the distribution system fluctuates throughout the day. During peak consumption periods, hydrants may flow less than what is indicated by the color they are painted. Furthermore, the possibility always exists that a hydrant has been painted the wrong color. Other possibilities abound for variations in water flow from hydrants. For example, operators may be

Table 7–3. *NFPA 291* hydrant markings

Hydrant classification	Color scheme	Hydrant capacity (gpm)
Class AA	Light blue	> 1,500
Class A	Green	1,000–1,499
Class B	Orange	500–999
Class C	Red	< 500

Source: NFPA, *NFPA 291: Recommended Practice for Water Flow Testing and Marking of Hydrants, 2022* (Quincy, MA: NFPA, 2021).

unaware of partially closed valves within the distribution system that restrict the water flow feeding the hydrant. Because of each of these variables, it is critically important for the pump operator to be proficient with calculating the available flow rate from a fire hydrant on the actual fireground.

Determining the available water from a hydrant on the fireground

Several methods exist for determining the volume of water that a fire hydrant is actually capable of providing in real time. The preferred option is known as the *percent drop method*, which is essentially an adaptation of the Hazen-Williams formula used for calculating the available flow from the system at a residual main pressure of 20 psi. In order to appropriately utilize this formula, the operator must know certain input data, including the static intake pressure coming from the hydrant, the residual intake pressure once water is flowing, and the flow rate leaving the supply pumper. The following formula can be used to determine the percentage drop in the intake pressure once water begins flowing from the supply pumper:

$$\%Drop = \frac{Static\ Intake\ Pressure - Residual\ Intake\ Pressure}{Static\ Intake\ Pressure} \times 100\%$$

Based on the percentage that the residual intake pressure drops, the remaining water availability can be determined:

- 0%–10% drop = Three times the current flow rate is available from the hydrant.
- 11%–15% drop = Two times the current flow rate is available from the hydrant.
- 16%–25% drop = One times the current flow rate is available from the hydrant.
- >25% drop = More water may be available from the hydrant, but not as much as the current flow rate.

This method is extremely accurate for determining the remaining water from a hydrant during an actual incident. The major disadvantage to this method, however, is that it requires the operator to quickly determine percentages in a highly stressful environment. The operator must then identify the relevant category that the percentage drop falls into in order to determine how much water can be expected from the hydrant. For these reasons, many firefighters forgo this method. In the grand scheme of things, the pump operator does not really care if the percent drop in the intake pressure is 5% or 9% since the percent drop in both of these instances falls within the 0% to 10% range. Instead of calculating the actual percent drop in the intake pressure on the fireground, the operator can utilize an available water flow conversion chart similar to the one in table 7–4.

Table 7–4. Available water flow conversion chart

	Three times available	Two times available	One times available	Less than current flow rate
Static pressure (psi)	0%–10% Drop (psi)	11%–15% Drop (psi)	16%–25% Drop (psi)	>25% Drop (psi)
200	200–180	178–170	168–150	148–1
190	190–171	169–162	160–143	141–1
180	180–162	160–153	151–135	133–1
170	170–153	151–145	143–128	126–1
160	160–144	142–136	134–120	118–1
150	150–135	134–128	126–113	111–1
140	140–126	125–119	118–105	104–1
130	130–117	116–111	109–98	96–1
120	120–108	107–102	101–90	89–1
110	110–99	98–94	92–83	81–1
100	100–90	89–85	84–75	74–1
95	95–86	85–81	80–71	70–1
90	90–81	80–77	76–68	67–1
85	85–77	76–72	71–64	63–1
80	80–72	71–68	67–60	59–1
75	75–68	67–64	63–56	56–1
70	70–63	62–60	59–53	52–1
65	65–59	58–55	55–49	48–1
60	60–54	53–51	50–45	44–1
55	55–50	49–47	46–41	40–1
50	50–45	45–43	42–38	37–1
45	45–41	40–38	38–34	33–1
40	40–36	36–34	34–30	29–1
35	35–32	31–30	29–26	25–1
30	30–27	27–26	25–23	22–1

The information in table 7-4 simplifies the process for determining the available water by eliminating the need for any complex calculations. Using this tool still requires the operator to note the static and residual intake pressures, as well as the flow rate being supplied. The leftmost column of the chart contains static intake pressures between 30 psi and 200 psi. The remaining columns represent the 0%–10%, 11%–15%, 16%–25%, and >25% drops in residual pressure for each given static pressure value.

After noting the static and residual intake pressures, the operator should find the row containing the static intake pressure value. The pump operator should then scroll across that particular row and locate the column that contains the residual intake pressure reading. Based on the column that the residual intake pressure reading falls under, the operator can determine how much more water the hydrant is capable of providing.

Using the available water conversion chart may indicate to the operator that the hydrant is capable of providing more water than the hydrant color indicates. Conversely, it may inform the operator that the hydrant is weaker than anticipated. A pump operator who makes any changes to the hydrant configuration, such as performing a heavy hydrant hookup, must remember to reference the conversion chart afterwards to determine the impact the configuration has on the water. Heavy hydrant hookups will usually result in the residual intake pressure rising when the flow rate remains constant. When referencing the chart, the original static intake pressure value is still used to determine which row the operator should scroll across.

During large flow operations, it is important to not only reference the residual intake pressure on the pumper, but also the residual hydrant pressure. Using the available water flow conversion chart, the operator may find that the residual intake pressure on the pumper has dropped beyond 25%, indicating that the operator cannot match the current flow rate. A look at the gauge placed on the hydrant, however, may reveal that the residual hydrant pressure is sufficient to match or exceed the current flow rate. In these situations, the operator must consider performing a heavy hydrant hookup, as mentioned previously. If a heavy hydrant hookup has already been completed and is in operation, it indicates that the main supplying the hydrant is strong. An additional hydrant should be used by a separate supply pumper in order to access more water from the main.

Heavy Hydrant Hookups

The single most effective way to increase the volume from a fire hydrant is to increase the outlet area of the hydrant. Heavy hydrant hookups involve hooking multiple supply lines to several hydrant outlets in order to maximize water availability. Prior to opening the main valve, the operator must first connect one or two hydrant gate valves to the 2½" side ports of the dry-barrel fire hydrant (fig. 7–2). This will allow additional supply lines to be connected without having to first shut down the hydrant.

When performing a heavy hydrant hookup, it is critically important for the operator to connect large diameter supply hose to the outlets being utilized. Many firefighters believe that 2½" or 3" supply line is sufficient for the connections made to the side ports. Since water travels through the path of least resistance, however, the friction loss within medium diameter hose (MDH) prevents any substantial increase in the volume from the hydrant. This is because the large diameter hose (LDH) connected to the steamer port has less resistance and therefore will deliver the majority of the water to the pumper.[15] The hose utilized for making the connections to each of the hydrant ports should be larger in diameter than the hydrant ports themselves. This will make the area of the hydrant outlet, not the hose diameter, the limiting factor when flowing large volumes of water.

Figure 7–2. At least one hydrant gate valve should be placed on the hydrant in order to expand the system if needed.

Double-tapped hydrant configurations

During a *double-tapped hydrant operation*, large diameter supply lines are connected from two separate hydrant ports to supply water into two separate pump intakes (fig. 7–3). Double tapping a hydrant should be considered the minimum action taken any time the operator connects to a fire hydrant. The operator must first connect the primary supply line from the supply pumper directly to the steamer port of the hydrant and then place a hydrant gate valve on one of the 2½" side ports. Although the flow rate through the steamer connection may be sufficient for the majority of fires, adding a gate valve to at least one side port provides the capability to maximize the hydrant capacity if necessary.

During large flow operations, it is possible for the friction loss in a single large diameter supply line to be significant enough to result in low residual intake pressures on the pumper's intake while still having sufficient residual hydrant pressure. This is an indication that more water is still available. After double tapping a hydrant and opening the second intake valve, the operator will typically see an increase in the residual intake pressure depending on the flow rate and diameter of the supply lines used. The reason for this is that when multiple hydrant outlets are utilized with LDH, the total flow rate from the hydrant is split between the openings. This ultimately reduces the amount of friction loss in the initial supply line, resulting in the

Figure 7–3. Double tapping should be considered the minimum action for increasing the available water from a hydrant.

increased residual intake pressure reading. LDH must be utilized for these operations to ensure that the area of each hydrant outlet is the restricting factor and not the hose diameter itself. Tables 7–5 and 7–6 both show how much water flows through each port of a double-tapped hydrant equipped with large diameter hose.[16]

Table 7–5. Double-tapped flow results at 1,000 gpm with LDH

Test description	Target flow from pumper (gpm)	Steamer port flow rate (gpm)	Side port flow rate (gpm)	Steamer port flow %	Side port flow %
4" to steamer and 4" to side port	1,065	645	420	61%	39%
5" to steamer and 5" to side port	1,065	745	320	70%	30%
5" to steamer and 4" to side port	1,065	750	315	70%	30%
6" to steamer and 4" to side port	1,065	760	305	71%	29%
6" to steamer and 5" to side port	1,065	755	310	71%	29%

Note: Percentage values are rounded to the nearest whole percentage.
Source: A. Soccodato, "Heavy Hydrant Hookups: Maximizing Hydrant Flows, Part 2," *Fire Apparatus and Emergency Equipment* 27, no. 8 (2022).

Table 7–6. Double-tapped flow results at 1,500 gpm with LDH

Test description	Target flow from pumper (gpm)	Steamer port flow rate (gpm)	Side port flow rate (gpm)	Steamer port flow %	Side port flow %
4" to steamer and 4" to side port	1,500	885	615	59%	41%
5" to steamer and 5" to side port	1,500	1,035	465	69%	31%
5" to steamer and 4" to side port	1,500	1,035	465	69%	31%
6" to steamer and 4" to side port	1,500	1,050	450	70%	30%
6" to steamer and 5" to side port	1,500	1,050	450	70%	30%

Note: Percentage values are rounded to the nearest whole percentage.
Source: A. Soccodato, "Heavy Hydrant Hookups: Maximizing Hydrant Flows, Part 2," *Fire Apparatus and Emergency Equipment* 27, no. 8 (2022).

A specific trend emerges when large diameter supply hose is used during double-tapped hydrant operations. As shown in tables 7–5 and 7–6, testing data indicate that when the supply line is larger than the steamer port connection, 70% of the total flow rate flows out of the line attached to the steamer port. The remaining 30% flows through the 2½" side port equipped with LDH. This division of labor between the two hydrant outlets results in a 50% reduction in the friction loss within the primary line connected to the steamer port.[17] The pump operator will see this as an increase in residual intake pressure on the supply pumper's master intake gauge.

The 70/30 split in water flow occurs whenever the supply line diameter is larger than the outlet to which it is connected. The only time this ratio does not occur is when a 4" supply line is connected to the steamer port. As evident from tables 7–5 and 7–6, hydrants that are double tapped with 4" supply lines have 60% of the total flow go through the line connected to the steamer port and 40% of the total flow exit through the side port. This is due to the fact that the 4" supply line is smaller than the 4½" outlet to which it is connected.[18] By double tapping a fire hydrant using 4" supply line, the friction loss in the 4" connected directly to the steamer port is reduced by approximately 63%. This explains why fire departments utilizing 4" hose for double-tapped operations report much larger increases in their residual intake pressures after opening the second line from the hydrant (fig. 7–4).

Figure 7–4. Double tapping the hydrant is extremely beneficial for departments that use 4" LDH.

It is critically important to remember that as the flow rate increases, the friction loss in each line also increases at an exponential rate. The benefit of using 5" or 6" LDH when double tapping the hydrant may not be evident at lower flows. This is due to the fact that the friction loss in these lines is minimal at lower flow rates. During high-volume operations, however, the use of this larger diameter hose will ultimately allow more of the hydrant's capacity into the pump with lower friction loss values. Firefighters must remember that double tapping a fire hydrant with LDH is the single most effective way to increase the efficiency of the supply operation when utilizing a hydrant.

Triple-tapped hydrant configurations

Triple-tapping hydrant operations entail connecting three large diameter supply lines to outlets on a fire hydrant (fig. 7–5). For maximum efficiency, it is recommended that two hydrant gate valves be attached to the 2½" side ports of the hydrant. This will allow the triple-tapped configuration to be built in stages as additional water is needed. Similar to the double-tapped configuration, it is critically important to ensure that only LDH is connected between the hydrant and the supply pumper.

Performing a triple-tapped hydrant operation is most beneficial if the apparatus is equipped with three large pump intakes. If the apparatus is only equipped with a driver's side and passenger's side main intake, the

Figure 7–5. A hydrant triple tapped with LDH

triple tap can still be achieved by using the 2½" auxiliary intake on the pumper (fig. 7–6), but it may not yield the same results. When the auxiliary intake is used for a triple-tapped operation, it is critically important that LDH continue to be used between the hydrant and the intake. This helps to reduce the friction loss within the system.

During triple-tapped hydrant operations, the total outlet area being used from the hydrant is larger than the area of the main valve. As previously discussed, this makes the hydrant itself the limiting factor with regard to maximizing the water source. The operator must understand that transitioning from a double-tapped configuration to a triple-tapped configuration will not typically result in a dramatic increase in residual intake pressure. This is especially true if 5" or 6" supply hose is used. Instead, this tactic allows higher flow rates to be achieved while maintaining a more comfortable residual intake pressure on the supply pumper. This is due to the fact that a smaller portion of the residual hydrant pressure is lost due to friction while traveling through each of the three supply lines.

A double-tapped hydrant configuration should be considered the minimum action taken when dealing with a hydrant. In contrast, a triple-tapped hydrant configuration should be performed when the operator is tasked with supplying heavy streams from an aerial device or when securing an additional water supply on a multiple alarm fire. Supplemental water

Figure 7–6. The auxiliary inlet can be used to triple tap a hydrant if a front or rear intake is not available. It is critically important to use LDH even when the auxiliary inlet is used.

supplies and heavy-caliber aerial streams will both require the full capacity of the hydrant. Triple tapping the hydrant with LDH is the only way to definitively maximize the volume from any single fire hydrant.

Heavy hydrant hookups and residual intake pressures

In order to understand the impact that heavy hydrant operations have on the available water for firefighting operations, we must evaluate the effect these operations have on residual hydrant and intake pressures. Table 7–7 lists the results of several tests that were conducted to evaluate this impact. The control test for this series utilized a single 5" supply line connected to the steamer port of the hydrant. A static intake pressure from the hydrant of 80 psi was recorded at the beginning of the tests. At a flow rate of 2,000 gpm through the single line setup, a residual intake pressure of 25 psi was recorded, but the hydrant showed a residual hydrant pressure of 52 psi.[19] This indicated that the hydrant had more water available, but that the friction loss within the single supply line would prevent the full potential from being achieved. Using the Hazen-Williams formula, this particular hydrant should be capable of providing roughly 3,000 gpm at a residual hydrant pressure of 20 psi.

Table 7–7. The effect of heavy hydrant hookups on residual intake pressures

Test configuration	Flow (gpm)	Residual intake pressure (psi)	Hydrant pressure (psi)
Single 5"	500	75	78
Single 5"	2,000	25	52
Double tapped w/ one 5", one 2.5"	2,000	25	52
Double tapped w/5"	2,000	40	52
Triple tapped w/5"	2,000	45	52
Double tapped w/5"	2,500	25	40
Triple tapped w/5"	2,500	25	40
Triple tapped w/5"	2,700	15	30

Source: A. Soccodato, "Heavy Hydrant Hookups: Maximizing Hydrant Flows, Part 2," *Fire Apparatus and Emergency Equipment* 27, no. 8 (2022).

The difference in the residual hydrant and intake pressures represents the friction loss in the system at 2,000 gpm. The effects of a double-tapped hydrant configuration on residual intake pressures were tested first, and initially a 2½" supply line was run between the hydrant side port and the pumper. After the hydrant gate valve and auxiliary intake were opened, it was noted that the residual intake pressure on the pumper remained at 25 psi, and the residual hydrant pressure remained at 52 psi while flowing 2,000 gpm.[20] This indicated that double tapping the hydrant with one 5" line off the steamer and one 2½" line on the side port provided absolutely no benefit to the operator (fig. 7–7). This is due to the fact that the 5" line off the steamer was the path of least resistance and thus was the easier path for water to travel through.

During the next test, the 2½" hose line was replaced with a 5" supply line on the hydrant side port. After making this change, the residual intake pressure on the pumper rose from 25 psi to 40 psi while flowing 2,000 gpm. The residual hydrant pressure remained at 52 psi.[21] This 60% increase in the residual intake pressure proved that double tapping the hydrant with LDH dramatically reduced the total friction loss between the hydrant and the supply pumper (fig. 7–8).

In order to evaluate the impact of triple tapping the hydrant, a third 5" supply line was added to the operation while the flow rate remained at 2,000 gpm. After opening the third intake, the residual intake pressure on the pumper rose from 40 psi to 45 psi, while the hydrant pressure remained at 52 psi.[22] Although this increase in pressure seems minimal, it is

Figure 7–7. Using MDH for a double-tapped hydrant operation provides almost no benefit to the operator.

Figure 7–8. Double tapping with 5" LDH resulted in a tremendous increase in the residual intake pressure.

important to recognize that after triple tapping the hydrant, the residual intake pressure and the residual hydrant pressure read within 7 psi of one another. The proximity of these values indicates that the total flow rate is split in such a way that it minimizes friction through the system.

The flow was eventually increased to 2,700 gpm from the triple-tapped hydrant configuration. At this flow rate, the residual intake pressure on the pumper was 15 psi, while the residual hydrant pressure was 30 psi. The increased difference between these two readings is due to the increase in the flow rate. Although the residual pressure on the master intake gauge read below 20 psi, the pressure at the hydrant remained above 20 psi, which made this operation compliant with the requirements set forth by most municipal water authorities and the AWWA.[23] The tests were concluded after reaching 2,700 gpm because the supply pumper reached the governed speed of its motor. It is important to recognize that the horsepower of the motor on the apparatus was the limiting factor in this operation, not the hydrant or the configuration of the supply hoses.

Double and triple tapping a hydrant with LDH will allow the friction loss in the system to be reduced as much as possible; however, friction will always exist at higher flow rates. This will likely cause the residual hydrant and intake pressure readings to differ from each other during high-flow operations. The pump operator must understand that when attempting to maximize the volume from a high-capacity hydrant, it is acceptable for the residual intake pressure to drop below 20 psi as long as the residual hydrant pressure remains above 20 psi. This makes it critically important that a pressure gauge of some sort be incorporated in the initial hydrant connection.

The operator must also understand that double and triple tapping a fire hydrant does not create water. Instead, these configurations allow the operator to maximize the water availability of high-volume hydrants by minimizing friction loss. This means that when low-volume hydrants are used, double and triple tapping will not magically cause more water to flow from them. Hydrant pressure gauges can help the operator make this determination regardless of the hydrant color. Due to the low friction loss in LDH at lower flow rates, the residual intake and hydrant pressure readings will likely closely match one another. The importance of this is that when the residual intake pressure reads 20 psi on the pumper, it likely will also read 20 psi at the hydrant (fig. 7–9). This would indicate that this is all the volume that the weak hydrant is able to safely provide.

Primary advantages of heavy hydrant hookups

The hydrant used for the tests listed in table 7–7 was capable of providing 2,000 gpm through a single 5" supply line, while the residual intake

Figure 7–9. When dealing with low-flow hydrants, the residual hydrant and residual intake pressures will read closer to one another. This is because there is minimal friction loss at low flow rates through LDH.

pressure remained at 25 psi. It can be argued that this flow rate is not necessary for the majority of fires that a department may respond to, so why go through the effort to perform a heavy hydrant configuration? The operator must remember that heavy hydrant operations increase the efficiency of the entire supply operation. There are two major advantages of using this tactic when operating from a hydrant:[24]

1. *Increased available water.* This is obviously the biggest advantage to performing a heavy hydrant hookup. Operators should remember that an increase in residual intake pressure directly correlates to more water being available from the hydrant through the supply configuration. Peak efficiency from high-volume hydrants is achieved when the residual intake pressure and the residual hydrant pressure readings closely match each other. The only way this can be achieved is by performing a heavy hydrant hookup to reduce the amount of friction between the hydrant and the pumper. While the configuration may be capable of supplying more volume than the fire scene requires, double or triple tapping the hydrant puts the operator in a position to increase the flow rate to the fire scene if it is needed.
2. *Reduced workload on the supply pumper.* Centrifugal fire pumps take advantage of pressure entering the pump from an outside source. This means that the higher the incoming pressure, the less work the centrifugal pump has to do in order to meet the pump discharge requirement for the supply operation. In the scenarios outlined in table 7–7, switching from a single 5" connection from the hydrant to a double-tapped 5" configuration resulted in a 15 psi increase in the residual intake pressure on the supply pumper. This increase in residual intake pressure meant that the supply pumper did not have to work as hard to provide the same pump discharge pressure. The result is a reduction in the needed engine revolutions per minute to provide the appropriate pump pressures and thus a reduction in the total amount of work done by the supply pumper. During incidents of longer duration, this will ultimately reduce the possibility of the apparatus running out of fuel, overheating, or having a mechanical issue.

Utilizing 6" supply hose for hydrant operations

Before we go any further in this section, it must be made clear that the author is not recommending that fire departments replace all the LDH on their engines with 6" hose. The 6" hose referenced here is soft suction hose with an internal diameter of 6" and a length of no more than 25'.[25] These sections of hose are intended to help maximize supply operations, especially when a fire hydrant is utilized.

The major advantage of using 6" hose for hydrant operations is the tremendous reduction in friction loss compared to traditional 4" or 5" supply lines. To put this into perspective, a single 6" supply line can move the same volume of water with the same amount of hydraulic loss as two 4" supply lines. This essentially means that when 6" hose is connected to a fire hydrant, it is the hydraulic equivalent of double tapping the hydrant with 4" hose (fig. 7–10). This should not be taken as a deterrent for performing heavy hydrant hookups. Rather, the operator should think of this as achieving the "most bang for the buck" with a single hydrant connection. Operators must still be prepared to perform a heavy hydrant operation to maximize the capability of a hydrant.

The use of 6" hose is extremely beneficial when utilizing a front intake on a pumper. As we have previously discussed, front intakes will incorporate greater internal plumbing losses than side intakes. This causes many fire departments to omit front intakes as an option when building pumpers. The front intake allows the operator unparalleled positioning options, however, especially in the urban environment. When these intakes are utilized with 6" hose, the operator is capable of positioning the supply engine to its best advantage, while also achieving impressive flow rates (fig. 7–11).

Figure 7–10. The 6" soft sleeve between the steamer port and the driver's intake is the hydraulic equivalent of two 4" supply lines.

Figure 7–11. The use of 6" soft sleeve greatly improves the flow capabilities of front intakes.

Because the friction loss in a 25' section of 6" hose is so low at higher flow rates, the only real cause of loss in the system is the front intake plumbing itself, which the operator has no ability to change. Using 6" hose to make a front intake connection allows more residual hydrant pressure to enter the front intake and be used to overcome the loss in the plumbing. The result is higher flow rates through the front intake compared to 4" or 5" LDH. Table 7–8 outlines the difference in residual intake pressures while flowing 1,000 gpm through the front intake of a particular pumper using 4" hose and 6" hose. The only variable changed during these tests was the hose diameter of the supply line connecting the hydrant to the front intake.[26]

Table 7–8 reveals that by simply switching from 4" hose on the front intake to 6" hose, the residual intake pressure on the apparatus increased by 10 psi. This increase in the residual intake pressure represents a 40% increase compared to what the 4" hose was capable of providing.[27] Some firefighters are under the impression that front intakes struggle to provide even 50% of the rated capacity of the pump. This is typically true when drafting from a front intake because there is a finite amount of atmospheric pressure available to overcome the lift, friction loss in the hard sleeve, and the friction loss in the front intake itself. When dealing with a high-volume hydrant with higher residual pressures, however, it is not unheard of to be able to flow 1,500 gpm through the front intake of a modern pumper using 6" hose.

Table 7–8. Effects of 6" hose on residual intake pressures

Configuration	Flow rate (gpm)	Residual intake pressure (psi)
4" connected to front intake	1,000	25
6" connected to front intake	1,000	35
6" connected to side intake w/ elbow	1,000	45

Source: A. Soccodato, "Maximizing Hydrant Flows Using 6-Inch Soft Suction," *Fire Apparatus & Emergency Equipment* 25, no. 10 (2020).

For departments using pumpers that are not equipped with front intakes, 6" hose can still be beneficial for hydrant operations using the side intakes of a pumper. Using this hose to connect directly to a side intake will obviously result in higher residual intake pressures due to the reduction in plumbing loss compared to a front intake. Positioning for a side intake can be difficult, however. Positioning too close to the hydrant will result in kinks in the supply line, while positioning further away will often block access for other apparatus. In these situations, the operator should elect to utilize a 90° elbow connected to the side intake of the pumper. This offers the same positioning advantages as a front intake while simultaneously allowing a 25' section of 6" hose to go directly to a main intake. It is important to note that this tactic can be used to increase the functionality of the supply pumper regardless of the diameter of the supply hose utilized.

The final row in table 7–8 highlights the impact that this supply configuration has on the residual intake pressure at a given flow rate. At a flow rate of 1,000 gpm, the residual intake pressure on the supply pumper was 45 psi, which represented the highest residual intake pressure out of the three configurations tested (fig. 7–12). It is important to recognize that the residual intake pressure recorded when the 6" supply line was connected directly to the driver's side intake using an elbow represented an 80% increase in the intake pressure compared to using 4" on the front intake.[28] The residual intake pressure increased roughly 29% after moving the 6" hose from the front intake of the rig to the driver's side intake using an elbow.

Four-Way Hydrant Valve Operations

All of the hydrant operations discussed to this point have assumed that a supply pumper is able to connect directly to the hydrant before the hydrant is turned on. While this may be possible for larger departments with a

Figure 7–12. These photos correlate to the information in table 7–8 and highlight the impact of running 6" soft sleeve directly to a pumper's main intake.

plethora of resources, smaller agencies may find it difficult to commit an engine to pump the hydrant at every fire. This could be due to limited staffing, long response times, or a differing fire department culture. Forward lays are often the method used to lay the supply line, and there could be 300' or more of supply line laid between the hydrant and the fire scene. Many fire departments will then connect the end of the supply line directly to the hydrant and use hydrant pressure alone to supply the scene.

The major disadvantage of using hydrant pressure to supply water to the scene is the inability of the hydrant to overcome the friction loss in the supply hose at higher flow rates, even if LDH is utilized. Example 7–1 showed a hydrant that is capable of flowing 2,591 gpm at a residual pressure of 20 psi. The friction loss in 5" hose at this flow rate is approximately 54 psi per 100', which is higher than the residual pressure of 20 psi, the pressure at which the hydrant is capable of supplying its rated volume. Fire departments that perform forward lays from the hydrant typically lay out more than 100' of supply line between the scene and the hydrant. This means that the available water to the scene using hydrant pressure alone will be significantly less.

For example, suppose the attack engine laid out 600' of 5" supply line from the hydrant used in example 7–1. The supply line is connected directly to the hydrant, with the other end connected to the attack engine's intake. We know that the static pressure of this hydrant is 70 psi and that once water starts flowing out of the attack pumper, the residual intake pressure and hydrant pressure will drop. Using the Hazen-Williams formula, we can calculate what the flow rate will be for a given residual hydrant pressure. Based on the flow rate for a given residual hydrant pressure, we can also determine the friction loss in the 600' of 5" supply line laid in the scenario. We then can determine what the residual intake pressure will be based on the pressure loss due to friction. Table 7–9 outlines the maximum flow rate that could be achieved using the residual hydrant pressure alone from this hypothetical hydrant to supply water through 600' of 5" hose line while maintaining a sufficient residual intake pressure.

Table 7–9 shows that the maximum volume this hydrant could supply through 600' of 5" hose while maintaining a sufficient residual intake pressure on the attack pumper is roughly 1,000 gpm. This would put the residual intake pressure on the attack engine somewhere around 10 psi, which is probably lower than most attack engine operators are comfortable using. It is important to note that even though the residual intake pressure on the attack pumper is relatively low, the residual pressure at the hydrant is still around 61 psi. The difference in these values is the friction loss between the hydrant and the pumper. The negative residual intake values represent an

Table 7–9. Maximum flow rate using hydrant pressure alone through 600' of 5" hose

Hydrant pressure (psi)	Flow (gpm)	Friction loss in 600' of 5" hose line (psi)	Residual intake pressure on attack engine (psi)
70	0	0	70
65	748	27	38
64	825	33	31
63	897	39	24
62	964	45	17
61	1,027	51	10
60	1,087	57	3
59	1,145	63	–4
58	1,200	69	–11

insufficient amount of residual intake pressure to keep the supply line from collapsing at flows greater than 1,087 gpm. This demonstrates that hydrant pressure alone is not sufficient to provide the rated capacity of the hydrant to the fire scene.

Table 7–10 shows what the flows and residual intake pressures would look like for the same scenario using 4" supply line instead. With 600' of 4" hose line laid between the hypothetical hydrant and the attack engine, we see that hydrant pressure alone could only realistically provide a maximum flow rate of around 650 gpm. This is significantly lower than what the hydrant is actually capable of providing. The solution in both of these situations is to place a pumper on the hydrant to boost the pressure and overcome the friction loss in the supply line. This will result in more of the capacity of the hydrant being delivered to the fire scene. If the supply line is connected to the hydrant, and the hydrant is opened prior to the arrival of the supply pumper, however, pumping the hydrant becomes more difficult. This is primarily because the hydrant will have to be shut down in order for the later-arriving supply pumper to make their connections to the hydrant. This will cause the fire scene to be without water while the transition takes place.

The four-way hydrant valve is the answer for departments that do not typically assign a supply pumper to pump the hydrant from the beginning of the incident and operate solely from hydrant pressure (fig. 7–13). These valves are connected to the hydrant at the beginning of the operation and allow the fire scene to establish a water supply using hydrant pressure. If the flow demands increase and cause more friction loss in the supply line

Table 7–10. Maximum flow rate using hydrant pressure alone through 600' of 4" hose

Hydrant pressure (psi)	Flow (gpm)	Friction loss in 600' of 4" hose line (psi)	Residual intake pressure on attack engine (psi)
70	0	0	70
69	314	12	57
68	456	25	43
67	568	39	28
66	663	53	13
65	748	67	–2

Figure 7–13. Four-way hydrant valves improve the efficiency of forward lay operations.

than the residual hydrant pressure can overcome, a later-arriving engine can be connected to the valve and boost the pressure. The major advantage the four-way valve provides is that it does not interrupt the supply of water to the fire scene while the supply engine makes the necessary connections to boost the pressure.

Several appliance manufacturers produce four-way hydrant valves, each with slightly different configurations. Some manufacturers build their valves in a side-by-side configuration, while others take an over-under approach to their design (figs. 7–14 and 7–15). But all the valves are designed to allow a later-arriving pumper to boost pressure and overcome friction loss in the initial supply line. Regardless of how these valves are configured, they have two positions that allow them to be used on the fireground. The first valve setting is known as the *straight through position* and refers to the setting that allows water to flow directly through the valve to the fire scene under hydrant pressure alone (fig. 7–16). The second valve setting is the *boost position*, which refers to the arrangement of the valve after the supply pumper has appropriately connected hoses to the valve and flowed water from the hydrant into the pumper to boost the pressure of the water being supplied to the fire scene (fig. 7–17).

Figure 7–14. Side-by-side four-way hydrant valve

Figure 7–15. Over-under four-way hydrant valve

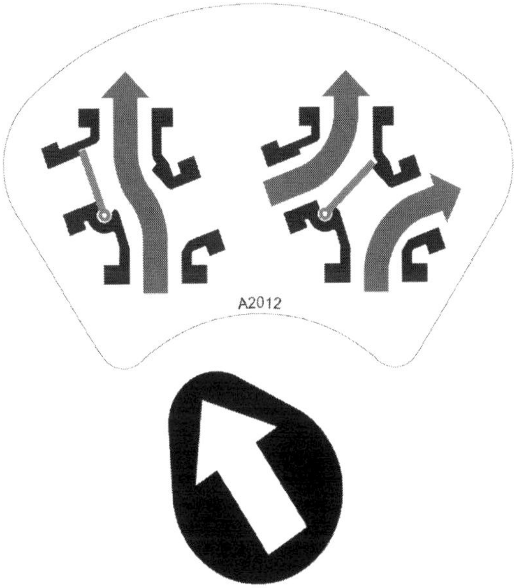

Figure 7–16. A four-way hydrant valve indicating that it is in the "straight through" position and is delivering water to the fire scene using hydrant pressure alone
(*Courtesy*: Task Force Tips)

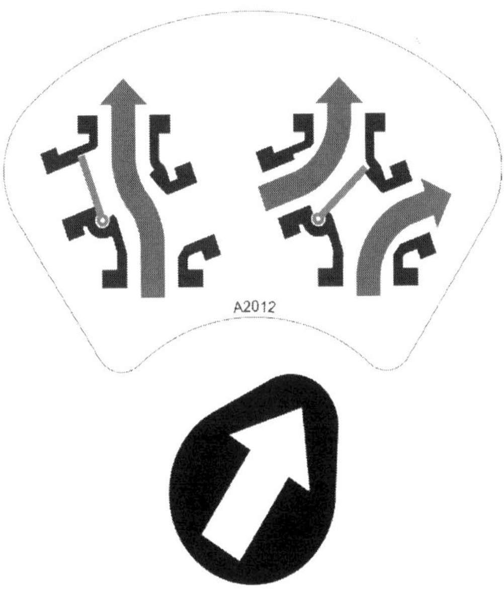

Figure 7–17. A four-way hydrant valve indicating that it is in the "boost" position
(*Courtesy*: Task Force Tips)

In order to utilize these valves appropriately, the operator must be familiar with the major components found on all four-way hydrant valves (fig. 7–18):

- *Hydrant connection.* Where the four-way valve connects to the steamer port of the fire hydrant.
- *"To Fire" outlet.* The connection outlet that is always sending water to the fire scene, regardless of whether the valve is in the straight through or boost position.
- *"To Supply Pumper" outlet.* This outlet is used to feed the supply pumper during a boost operation. A supply line is connected from this port to an intake on the supply pumper.
- *"From Supply Pumper" inlet.* A line from the supply pumper is connected to this port, and the other end is connected to the pumper's large diameter discharge. This connection allows the supply pumper to boost the pressure and overcome the friction loss in the supply line.
- *Crank/valve handle.* This mechanism directs water out of the four-way valve. In the straight through position, the crank/valve handle is positioned to only allow water to exit from the "To Fire" outlet. When the crank/valve handle is manipulated to the boost position, it allows water to flow from the "To Fire" outlet and the "To Supply Pumper" outlet simultaneously.

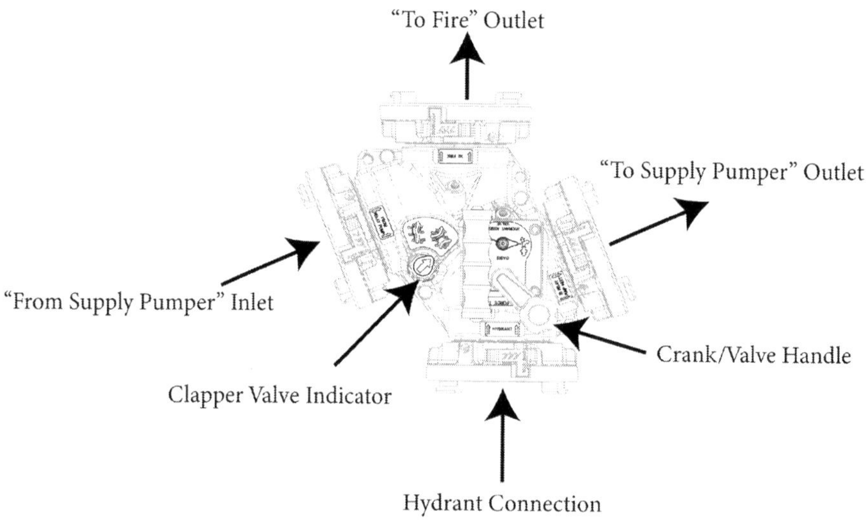

Figure 7–18. Components of a typical four-way valve
(*Courtesy*: Task Force Tips)

- *Clapper valve.* The role of the clapper valve is to divert the flow of water between the two sides of the four-way valve. In the straight through position, hydrant pressure acts on the clapper valve to keep water flowing only out of the "To Fire" port. During a boost operation, once the pressure from the supply pumper exceeds the residual hydrant pressure, the clapper valve opens fully within the four-way valve. This causes all the water from the hydrant to flow out of the "To Supply Pumper" outlet and feed the supply pumper. The supply pumper then adds pressure to the water and sends it through the "From Supply Pumper" inlet and on to the fire scene.

Traditional four-way valve operations

Whether the initial engine company lays the supply line, or a second-arriving pumper performs a booster back up, typically the first step in using a four-way valve is to leave the appliance at the hydrant and perform a forward lay. The valve is then connected directly to the hydrant, and the supply line is connected to the "To Fire" outlet. Once the attack engine is ready for water, the hydrant can be opened, and water can be sent to the fire scene. As long as the hydrant pressure is capable of supplying the needed fire flow and overcoming the friction loss in the supply line, nothing further needs to be done with the four-way valve (fig. 7–19).

Firefighters should think of the four-way valve as an insurance policy, i.e., it is there if it is needed, but it may not be used at every fire scene. If the flow rate at the scene increases to a point where the hydrant does not have enough residual pressure to overcome the friction loss in the supply line, a

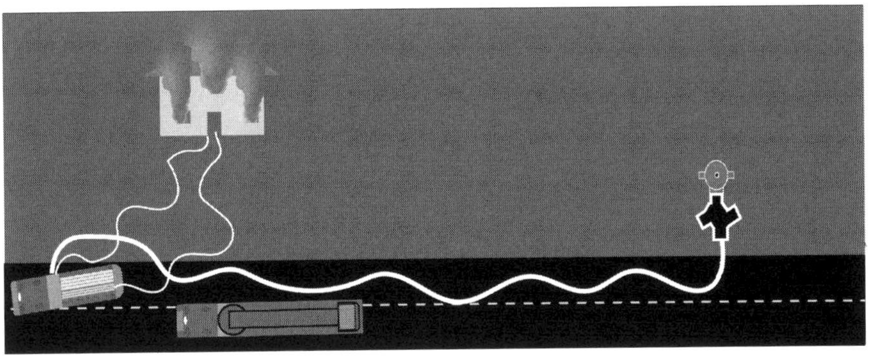

Figure 7–19. The attack engine performed a forward lay from the hydrant using a four-way valve.

later-arriving pumper can be used to connect to the four-way valve. This pumper's first responsibility is to position appropriately at the hydrant and make the necessary connections. After the supply lines are connected between the four-way valve and the pumper, the operator of the supply engine should slowly open the crank/valve handle to allow water from the hydrant to flow from the "To Supply Pumper" outlet. Once this line is charged, the operator of the supply engine must open the intake to receive water from the hydrant.

At this point, the four-way valve is essentially acting as a giant wye and supplying water to the fire scene and to the supply pumper simultaneously. The discharge between the supply pumper and the four-way valve should then be charged, and the throttle should be increased to build the necessary pressure. Once the discharge pressure exceeds the residual pressure from the hydrant, the clapper valve is moved completely to the boost position. This causes all the water exiting the steamer on the hydrant to enter the supply pumper's intake. The supply pumper adds pressure to the water entering the pump before discharging it to the scene. As the supply engine builds pressure through the four-way valve, the attack engine should see a corresponding increase in their residual intake pressure. Using this appliance allows the supply pumper to make this transition seamlessly without interrupting the flow of water to the fire scene (fig. 7–20).

Firefighters must recognize that these appliances do have some associated appliance loss. The amount of loss within these valves varies directly with the style of valve utilized and the volume of water flowing through the valve. In the boost position, there are essentially two sides of the four-way valve: the intake side feeding the supply engine and the discharge side where

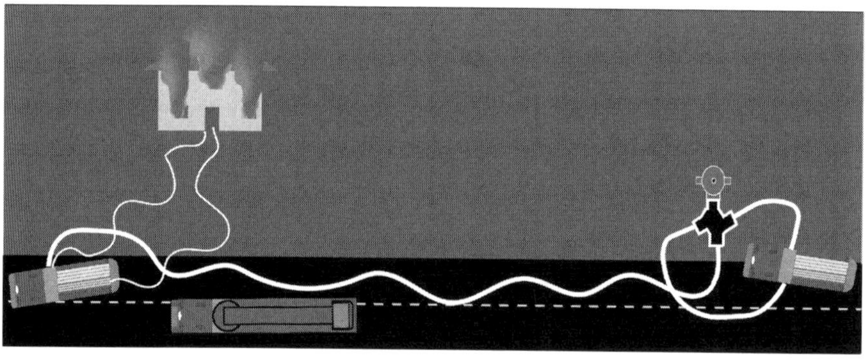

Figure 7–20. In the event that the pressure needs to be boosted, a later-arriving pumper can connect to the four-way valve.

the boosted water flows through to the fire scene. Technically speaking, loss occurs on both of these sides. The supply engine is capable of compensating for the appliance loss on the discharge side of the valve by building more pressure. The associated loss on the intake side of the valve can have a significant impact on how much water is capable of flowing from the hydrant and into the supply pumper, however.

For this reason, it is critically important that firefighters ensure that a hydrant gate valve is placed on the hydrant during any four-way hydrant valve operation. This will allow the supply operator to easily double tap the hydrant, which reduces the total volume of water that must flow through the intake side of the four-way valve (fig. 7–21). By doing this, the operator can effectively minimize the total amount of appliance loss on the intake side of the valve. The immediate result will be a higher residual intake pressure on the supply engine for a given flow rate. It should be noted that due to their design, the over-under style four-way valves tend to have a minimal amount of appliance loss when operating in the boost position. Departments using this style of valve should still place at least one hydrant gate valve on a 2½" side port during four-way valve operations in order to maximize their capabilities.

Once again, it must be understood that four-way hydrant valves do not create water. Instead, these valves allow a later-arriving engine to deliver more of the rated capacity of a hydrant to the fire scene by overcoming

Figure 7–21. It is critically important that gate valves be placed on the hydrant when using a side-by-side style four-way valve.

friction loss in the supply hose. If a four-way valve is connected to a low-flow hydrant, using it in the traditional manner will not result in achieving a flow greater than the rated capacity of the hydrant. Operators must recognize this before committing resources to pump a hydrant that simply does not have sufficient water to be effective.

Alternative four-way valve operations

Although four-way hydrant valves are traditionally associated with forward lay operations, they can be used for other water supply scenarios. In order to use these appliances in alternative configurations, it is critically important for pump operators to be highly familiar with the specific valves used by their departments. Failure to do so could severely disrupt the supply of water to the fire scene.

In situations where the operator does not have enough hydrant gate valves to perform a heavy hydrant hookup, the four-way valve can be used as a wye to help split the flow. This configuration requires that the valve be connected to the steamer port normally. A supply line is then connected to the "To Fire" port with the other end going directly into the supply pumper. Once the hydrant is opened, the flow of water will go directly into the supply engine, rather than to the fire scene. When the operator elects to double tap the hydrant, the second supply line is connected to the "To Supply Pumper" port on the four-way valve. The other end of this supply line is connected to a second intake on the supply pumper. By manipulating the crank/valve handle, the operator allows the water flowing from the steamer port to be split between the two outlets on the four-way valve.

The flow rate that can be achieved in this configuration will depend on the style and internal waterway of the four-way valve utilized. Regardless, a hydrant gate valve should still be attached to one of the side ports of the fire hydrant. This will allow a third supply line to be connected from the hydrant to the supply pumper, which will reduce the total amount of friction in the system. There is the potential for a fourth LDH supply line to be connected to the engine if the operator is able to connect two hydrant gate valves and the four-way hydrant valve to the hydrant from the beginning of the operation (fig. 7–22).

Certain fire scenes may require the implementation of a *split lay* in order to establish a water supply at the fire scene. During a split lay, the initial engine will stop, drop their supply line, and perform a forward lay to the fire scene. The next engine to arrive will stop where the first engine began

Figure 7–22. Using this four-way valve as a wye and using both side ports of the hydrant allowed this pumper to make four connections to this hydrant using LDH.

their layout, connect their supply line to it, and reverse out toward the water source. The distance between the source and the fire scene will dictate how much hose separates the two pumpers. Depending on the flow rate required at the fire scene, the supply engine could struggle to overcome the friction loss in the hose layout and may find it difficult to provide the necessary volume to the fire scene. The solution would be to place a relay pumper somewhere in the middle of the layout. If the third engine arrives after the operation is already placed in service, however, the supply of water will have to be shut down in order to add this pumper to the system.

Four-way hydrant valves can be used in these situations to efficiently incorporate the third pumper by using the appliance as an in-line relay valve. To use the valve in this fashion, the initial engine should leave a four-way valve at the point where they begin their forward lay. The supply line from the attack engine should be connected to the "To Fire" port. When the second engine arrives, they will connect their supply line to the portion of the valve that would normally connect to the steamer port of the hydrant. This may require the use of an adapter depending on the hose threads or connections used. The second engine then performs a reverse lay and continues to the water source. The supply of water is then initiated as outlined above. If a third engine needs to be added to the operation as a relay pumper, it can be accomplished using the four-way valve. This engine simply positions at the valve and connects to it the same way they would if the valve were connected to a hydrant. The principles for boosting pressure in this configuration are the same as for traditional four-way hydrant valve operations.

Although four-way valves can do nothing to increase the flow capacity of a low-flow hydrant, they can be used to help transition the operation to a stronger water supply. Often when an engine company performs a forward lay, it does so from the hydrant closest to the scene, not necessarily the one with the highest flow rate. The potential thus exists for the attack engine to perform a forward lay from a weak or even dead hydrant.

In these scenarios, the second engine should stop at the initial hydrant and prepare for a reverse layout. The second engine should then connect their supply line directly to the "From Supply Pumper" port on the four-way valve and reverse out toward a stronger water source. Once at the source, the operator should make the appropriate connections and prepare to send water to the attack engine. As the supply line is charged and pressure is built, the higher pressure of the water coming from the supply pumper will cause the internal clapper valve to move. This shifts the supply operation to the reliable water source (fig. 7–23).

Some fire departments elect to use their four-way hydrant valve for rural hitch operations in lieu of a separate clappered siamese. This reduces the number of appliances that must be carried by an engine company (fig. 7–24). Using the four-way valve for these operations may also require the use of

Figure 7–23. A four-way hydrant valve being used to overcome a dead hydrant

Figure 7–24. This department uses their four-way valve for rural hitch operations.

the same adapter mentioned for in-line relay operations to allow the necessary hose connections to be made. Successful use of this appliance for the rural hitch requires that firefighters understand where to appropriately connect their supply lines for the operation. Connecting to the wrong port when using the valve for the rural hitch could prevent the supply operation from unfolding as outlined in chapter 5.

Other Advanced Hydrant Operations

The previous sections discussed tactics that can be utilized for the majority of hydrants encountered across the United States. This section will outline more advanced tactics that can be utilized for maximizing flow capabilities while operating from a hydrant. Not all of these tactics may be applicable to every fire department. When used appropriately, however, these options will not only increase the volume of water that can be achieved from the hydrant but will also simplify the water supply operation as a whole.

Pressure-rated hard sleeves

Firefighters who have been in the fire service for any length of time have probably seen pictures of engines connected to fire hydrants utilizing sections of hard sleeve (fig. 7–25). The common misconception with these operations is that the fire department pumper is being used to "draft" from the municipal water supply system. This is impossible since municipal water distribution systems are pressurized sources. So why do fire departments utilize this tactic on today's fireground?

Prior to going any further with this discussion, firefighters must first determine the type of hard suction hose utilized by their fire department. Most sections of hard suction hose purchased today are only intended to be used for drafting operations. Accidentally using these sections of hard sleeve for hydrant operations will result in a catastrophic failure of the hose when the hydrant is opened. *This could injure or kill firefighters operating in the vicinity.* Before connecting any type of hard sleeve to a fire hydrant, firefighters must first ensure that the hard sleeve they are using is rated for positive pressure.

Departments that are using pressure-rated hard sleeve typically purchase it with a 6" diameter. This allows the hose to be utilized for both drafting and hydrant operations. One of the major advantages of using this hose for hydrant operations is the tremendous reduction in friction loss compared to 4" and 5" soft suction hose. The other major advantage is the fact that hard suction hose requires no pressure to remain charged. This means that all of the residual hydrant pressure is used to overcome the minimal amount of friction in the hard sleeve and to push water into the pump. These two

Figure 7–25. Pressure-rated hard sleeve can tremendously increase the efficiency of a hydrant operation.

characteristics allow the supply pumper to achieve more of the rated capacity of the hydrant.

One of the major disadvantages of pressure-rated hard sleeve is that it is not nearly as flexible as "drafting-only" hard sleeve. In addition, this hard sleeve also tends to be heavier compared to drafting-only sleeve. These characteristics can make it more challenging to connect this hose for hydrant operations. Positioning the apparatus appropriately becomes paramount when pressure-rated hard sleeve is used for hydrant operations. If the engine positions too near the hydrant, making the necessary bends to connect the hard sleeve to the steamer outlet and pumper inlet could be nearly impossible (fig. 7–26).

In an effort to make these connections easier, the fire department should consider utilizing a 90° elbow. The major advantage to using the elbow is that it eliminates the need to make one of the hard bends in the hose (fig. 7–27). The elbow can be placed at either end of the hard sleeve when making the connection. When making these connections, it is easiest for the engine to stop short of the hydrant and first gather the necessary equipment. Just like dealing with a dry hydrant in the rural setting, the operator should connect two sections of pressure-rated hard sleeve together. Next, the hard sleeve should be connected to the steamer port of the hydrant using the appropriate adapters. The operator should then bend the entire hose assembly in the direction of the pumper and connect the elbow to the

Figure 7–26. Pressure-rated hard sleeve can be quite difficult to bend into position.

Figure 7–27. Using the elbow made hooking to this hydrant a one-person operation.

other end. Now the pumper can be repositioned to allow the side intake of the apparatus to align with the hard sleeve and elbow.

Although this hose alone tremendously reduces the friction loss between the hydrant and the pumper, it is still imperative that hydrant gate valves be placed on any additional side ports. This will allow the operator to completely maximize the flow rate from a single hydrant if required. Even if pressure-rated hard sleeve is used, the need to double or triple tap the hydrant could arise if the hydrant is on an extremely strong main. It is important to note that if the fire department decides to invest in this type of hard sleeve for these operations, they should first ensure that their intake valves have large waterways. Intakes with 3½" waterways may render this tactic useless due to the associated friction loss at higher flow rates. Such waterways will greatly increase the need for engine companies to double tap and triple tap their hydrants in order to split the total flow rate between the intakes on the pumper.

The pump operator must recognize that the use of pressure-rated hard sleeve for hydrant operations is not a tactic intended for initial supply operations. This is due to the fact that implementing this type of operation takes time. Pumpers establishing initial water supplies should be focused on rapidly making their connections, preferably using 6" soft sleeve, and delivering water to the fire scene. The utilization of pressure-rated hard sleeve should be reserved for multiple alarm fires where secondary and tertiary water supplies are being established. It is the standard practice in some large cities, such as Chicago, to implement this tactic for supply engines operating on multiple alarm fires. This is due to the fact that the water system in the City of Chicago is considered a high-volume, low-pressure water distribution system.[29] Implementing this tactic allows Chicago engine companies to access more of the water available from the city's water distribution system.

Overcoming blocked hydrants

When operating in the urban environment, it is not uncommon to come across a fire hydrant that is blocked due to a parked car. This scenario has been portrayed in pop culture in movies such as *Backdraft*, with Firefighter John "Axe" Adcox smashing out the windows of a car blocking Engine 17's access to the hydrant.[30] Although most firefighters dream of the opportunity to replicate the actions of Axe in real life, they should also consider whether or not this is the best course of action. Aside from the property damage that this causes, firefighters must also be mindful of the kinks that will likely result from running supply hose through the broken windows of the vehicle. This section will discuss alternative options for dealing with blocked hydrants that not only minimize damage, but also provide respectable flow rates.

Some vehicles may be parked so close to the hydrant that there is absolutely no prospect of utilizing the steamer port. The best option in these situations is to double tap the hydrant using the two 2½" side ports. This prevents kinks in the supply line by allowing each hose line to be directed around the vehicle obstructing the hydrant. It is absolutely critical that LDH be used in these situations to help minimize the friction loss in the system. Many firefighters believe that this configuration is incapable of providing sufficient flow rates and therefore should not be utilized. Table 7–11 summarizes the results of several tests conducted to see what types of flow rates could be achieved from a high-volume hydrant that had its steamer port blocked by a vehicle (fig. 7–28).

These tests proved that the supply pumper was capable of achieving a flow rate of 2,000 gpm just by double tapping the side ports of the blocked hydrant with 5" hose. The pump operator must remember that the only hose that should be utilized for these operations is LDH supply line. It is also worth noting that the reason for concluding these tests at a flow rate of 2,000 gpm was not because the residual intake pressure was at 20 psi.

Table 7–11. Flow rates through 2½" side ports of a blocked hydrant with different hose configurations

Test configuration	Residual hydrant pressure (psi)	Residual intake pressure (psi)	Flow rate (gpm)
Single 3"	100	5	850
Dual 3"	96	45	1,188
Single 5"	94	55	1,250
Dual 5"	94	85	1,250
Dual 5"	70	20	2,000

Figure 7–28. Overcoming a blocked hydrant by double tapping the side ports with LDH

Instead, these tests halted due to the fact that the supply pumper reached the governed speed of its motor.

The use of the 2½" side ports in this fashion is not limited only to blocked hydrant scenarios. The pump operator should also consider this tactic if the nut on the steamer cap is sheared off or the cap itself is frozen to the hydrant. In other instances, the operator may find that the nut on the steamer cap is rounded or stripped. Ideally, the operator has a pipe wrench in the hydrant bag to overcome this issue. In the event that a pipe wrench is not available, however, the operator should consider double tapping the hydrant using both of the 2½" side ports.

If the vehicle is parked further away from the steamer port, another option is to simply use a 90° elbow on the steamer port. This will help to direct the supply line in the direction of the supply engine and is arguably the easiest of all the options for overcoming this obstacle. The operator must still remember to place gate valves on the side port of the hydrant in order to perform a heavy hydrant hookup. A four-way hydrant valve can also be used in a similar fashion when an elbow is not available to the operator. This can only be done if there is sufficient space between the car and the steamer port of the hydrant to fit the four-way valve. If this is not the case, the operator should elect to use the side ports of the hydrant.

Both side-by-side and over-under four-way valves can be used for blocked hydrant scenarios, but the over-under four-way valves tend to make this operation a bit easier. This is because when the hydrant is blocked, the appliance is still connected to the steamer port in its traditional fashion. After connecting the over-under four-way valve to the steamer port, the operator simply runs a supply line from their intake and connects it to the "To Fire" connection on the valve. When the hydrant is opened, water will flow directly into the supply pumper.

Side-by-side four-way valves can also be used for overcoming a blocked hydrant. The simplest option involves connecting the "From Supply Pumper" port directly to the hydrant steamer outlet. A supply line from the intake of the pumper can then be connected to the "To Fire" port on the four-way valve. Once the hydrant is opened, the clapper valve in the appliance will move and only allow water to flow out of the "To Fire" outlet. The pump operator is essentially using the four-way valve as an elbow in this arrangement (fig. 7–29).

Double steamer hydrants

Some localities have fire hydrants installed that are equipped with two steamer outlets (fig. 7–30). Double steamer configurations are found on large water mains and will typically provide more water than a single pumper is capable of moving. This is primarily because the motor of a single pumper will likely run out of horsepower before reaching the rated capacity of the hydrant. In order to maximize the available water from these special hydrants, firefighters must utilize more advanced water supply practices.

Figure 7–29. Using a four-way hydrant valve to overcome a blocked hydrant

Figure 7–30. Double steamer hydrants are found on extremely strong mains.

Tests conducted with the Johnson City (TN) Fire Department were performed to determine the best configuration for maximizing the flow capability of double steamer hydrants. Table 7–12 summarizes the results of three test configurations that were evaluated. The initial test performed was to determine what a single engine could flow using a single port on the hydrant. This configuration resulted in a flow rate through the pumper of 2,500 gpm with a residual intake pressure of 25 psi and a value of 1,660 rpm for the engine. This single engine then connected a supply line to each of the steamer ports of the hydrant and ran them to separate intakes on the pumper (fig. 7–31). While flowing 2,500 gpm, the residual intake pressure increased to 45 psi, and the engine speed dropped to 1,505 rpm.

The next configuration involved placing two engines on the double steamer hydrant. Each engine had a single LDH supply line connected to a steamer of the hydrant. The two pumpers were then connected to one another via an intake-to-intake operation. Building the system in this water supply configuration essentially created a looped system that allowed both pumpers to access the maximum volume of water this hydrant was capable of providing (fig. 7–32). After establishing this operation, a flow rate of 3,500 gpm was able to be achieved with a residual intake pressure on both pumpers of 5 psi. It is notable that by configuring the apparatus in this fashion, the flow rate was able to be increased by 1,000 gpm, while both engines operated below their governed engine speeds.

Municipalities may not always install double steamer hydrants on exceptionally large or strong water mains. Reasons for this vary, but the operator can maximize a strong three-outlet hydrant using some of the same

Table 7–12. Double steamer hydrant flow tests

	Single engine, single connection	Single engine, double tapped	Two engines, dual pumping
Total flow rate (gpm)	2,500	2,500	3,500
Engine 8 flow rate (gpm)	2,500	2,500	2,500
Engine 8 residual intake pressure (psi)	25	45	5
Engine 8 (rpm)	1,660	1,505	1,600
Engine 6 flow rate (gpm)	—	—	1,000
Engine 6 residual intake pressure (psi)	—	—	5
Engine 6 (rpm)	—	—	1,500

Figure 7–31. Johnson City Engine 8 double tapped to this double steamer hydrant.

Figure 7–32. By hooking two pumpers to this double steamer hydrant, a flow rate of 3,500 gpm was achieved.

principles previously discussed. When dealing with a known three-outlet hydrant of high capacity, two pumpers should be utilized to maximize the volume from the source. This will require the initial engine to be equipped with at least three large intakes. The original supply engine should fully dress the hydrant by placing two gate valves on the 2½" side ports. Next, the supply engine should double tap the hydrant using the steamer port and one of the side ports.

Later-arriving supply pumpers should position as near to the hydrant as possible to keep their supply lengths short. The second supply engine should initially hook one LDH supply line between one of their intakes and the unused side port of the hydrant. After completing this, the second pumper should perform a dual pumping operation with the original supply engine. This will allow a looped supply system to be created similar to the one outlined for double steamer hydrant operations. While using this configuration on a traditional hydrant may not be as efficient as it is on a double steamer hydrant, it still allows for the use of two diesel motors to work together to provide a higher flow rate to the fire scene (fig. 7–33).

Hydrants with excessively high pressures

Depending on the design of the water distribution system, certain localities may have to deal with hydrants that provide extremely high pressures. This typically occurs when dealing with industrial manufacturing facilities that are equipped with their own private water system. In these facilities, the same lines that provide water for a particular manufacturing process may also deliver water for fire protection. Certain manufacturing processes could require higher pressures in the distribution system that result in higher hydrant pressures as well. While this might initially appear advantageous, high hydrant pressures can potentially cause problems for firefighting

Figure 7–33. These two pumpers are both connected to this strong hydrant and are performing an intake-to-intake operation with each other. This allows both pumpers to take advantage of the volume in this hydrant.

operations. These high pressures have the potential to burst hose lengths and overpressurize the attack lines deployed off the attack engine.

The traditional solution to this problem is to simply set the intake relief valve to a lower setting to prevent high pressures from entering the pump. *NFPA 1900* states that when the intake relief valve is set to 125 psi, the valve must not allow the intake pressure to rise more than 60 psi at the device inlet while flowing at least 150 gpm.[31] If the hydrant has exceptionally high pressures, the intake relief valve may not be capable of flowing enough water to prevent overpressurization. The other obvious disadvantage to relying on the intake relief valve is that it will be dumping water around the pump panel for the entire incident. This may not be an issue in the summer, but it can be a significant problem during the winter.

An untraditional solution to this problem involves the use of a four-way valve in conjunction with a drafting operation. In order for this operation to be most efficient, it is preferred that the four-way valve be equipped with a crank/valve handle that can be manipulated to an "Off" position. In the "Off" position, the valve essentially prevents water from the hydrant from entering the four-way valve and exiting out of the "To Fire" or "To Supply Pumper" outlets.

The first step when establishing a water supply in these situations is for a supply line to be laid between the fire scene and the high-pressure hydrant. The four-way valve should be connected to the hydrant, and the other end of the supply line should be connected to the attack engine. It is critically important that hydrant gate valves also be attached to the hydrant as well. Once the attack engine is ready for water, the hydrant is opened, and water is supplied just as it would be during a traditional four-way valve operation. At this point, the attack engine has water, but they are dealing with excessive intake pressure.

After the hydrant is opened and the fire scene is supplied with water, the supply engine can begin building a system to reduce the amount of intake pressure on the attack pumper. This will require the supply engine to perform a drafting operation. The first step will be for a dump tank to be set up in the vicinity of the hydrant. Next, lines from the hydrant gate valves on the hydrant should be run and secured to the dump tank. The gate valves should be slowly opened to begin filling the dump tank once the hose lines are properly in place. The supply engine must then connect a supply line from their large diameter discharge to the "From Supply Pumper" port on the four-way valve. After completing that connection, the supply operator should prepare for and achieve a draft out of the dump tank.

The operator is now ready to transition from a high-pressure hydrant supply operation to a low-pressure drafting operation. The first step in this transition is for the operator to charge the discharge feeding the four-way valve. Since the pressure from the hydrant is extremely high, the clapper valve in the four-way valve will remain closed, allowing water from the hydrant to continue supplying the scene. At this point, the operator should manipulate the crank/valve handle to the "Off" position on the four-way valve. By doing this, the high-pressure hydrant is no longer supplying water through the supply line to the fire scene. This will immediately cause the internal clapper valve to open, which will send the water being drafted by the supply engine to the fire scene (fig. 7–34).

The major advantage of performing this type of operation is that the supply pumper is in complete control of the pressure feeding the attack engine. This will help to minimize any overpressurization issues at the attack engine. After making the transition, the supply operator must be mindful to open the gate valves of the hydrant to continuously top off the dump tank. This is why it is so important to have a four-way valve that allows for the hydrant to remain open but not send water to the fire scene. The beauty of this tactic is that at any point in time, the operation can be transitioned back to the high-pressure hydrant if needed.

Figure 7–34. This high-pressure hydrant is being used to fill the dump tank. The supply pumper is drafting from the dump tank and pumping through the four-way valve to provide the fire scene with a manageable intake pressure.

Study Questions

1. Fire hydrants are rated to flow their maximum volume at a residual pressure of:

   a. 10 psi
   b. 20 psi
   c. 30 psi
   d. 50 psi

2. You are operating the supply pumper connected to a fire hydrant. The static intake pressure from the hydrant was 80 psi. The residual intake pressure while flowing 500 gpm is 65 psi. How much additional water is available from this hydrant?

   a. 1,500 gpm
   b. 1,000 gpm
   c. 500 gpm
   d. Less than 500 gpm

3. Which of the following should be considered the minimum action taken when hooking to a hydrant?

   a. Only connect to the steamer port
   b. Double tapping the hydrant
   c. Triple tapping the hydrant
   d. Only using one 2½" side port

4. Triple tapping the hydrant should be done in which of the following situations?

   a. Low-volume hydrants
   b. Single line fires
   c. High-volume hydrants
   d. Only if medium diameter hose is used

5. Which of the following is true regarding four-way hydrant valves?

   a. An engine must pump through it whenever it is hooked to the hydrant.
   b. They allow a later-arriving engine to hook to the valve and boost pressure without first shutting down the hydrant.

c. They require the hydrant to be shut down before the engine can hook to the hydrant.
d. They create additional water from the hydrant.

6. Which of the following is an alternative use for a four-way hydrant valve?

a. Overcoming a blocked hydrant
b. Using it as an in-line relay valve
c. Overcoming a dead/low-flow hydrant
d. All of the above

7. What must the operator ensure prior to hooking hard sleeve up to a fire hydrant?

a. That it is rated for positive pressure
b. That the municipal system can manage a vacuum
c. That the engine is positioned close to the hydrant
d. All of the above

8. Double steamer hydrants:

a. Are found on weak mains
b. Do not require a supply engine to pump them
c. Are capable of providing more water than a single engine is capable of moving
d. Are intended for drafting operations

9. Two engines operating from a double steamer hydrant can maximize their configuration by performing what operation?

a. Tandem pumping operation
b. Dual pumping operation
c. Relay operation
d. Drafting operation

10. Which of the following is the best option for overcoming a hydrant blocked by a vehicle?

a. Run the supply line through the broken windows of the vehicle.
b. Hook a single 2½" line to one of the 2½" side ports.
c. Double tap the hydrant using 2½" hose connected to both side ports.
d. Double tap the hydrant using LDH connected to both side ports.

8

WATER SUPPLY FOR AERIAL APPARATUS

Aerial apparatus are some of the most impressive pieces of modern firefighting machinery. Advances in aerial technology have made aerial apparatus capable of performing a multitude of tasks on the fireground. During large-scale fires, modern aerial apparatus are typically used for their ability to provide heavy-caliber elevated master streams. When used correctly, the streams produced by aerial apparatus have the potential to extinguish large volumes of fire. Unfortunately, many organizations fail to utilize these powerful firefighting machines to their full potential.

Often firefighters think of elevated master stream operations as simply hooking a supply line to the aerial inlet and opening a valve. Although this will probably result in water flowing out the tip of the ladder, it will likely lack efficiency. In order to appropriately supply water to any type of aerial apparatus, pump operators must have an understanding of the components of the particular aerial apparatus they are feeding. It is even more important for the aerial operators to have detailed knowledge of the firefighting capabilities of the particular aerial apparatus they are assigned to operate.

This chapter will discuss the components found on most modern aerial apparatus that are used for elevated master stream operations. Variables that the supply pump operator must take into consideration while supplying large volumes of water to the aerial will also be discussed. It is vital for the reader to understand that aerial apparatus are extremely specialized and unique pieces of firefighting equipment. This chapter is intended to broadly touch on the capabilities of most aerial apparatus currently on the market. After reading this chapter, it is critically important for readers to review the operator's manual of the specific aerial apparatus used by their department. Failure to follow the apparatus manufacturer's recommendations for a specific aerial apparatus could have catastrophic consequences.

Types of Aerial Apparatus

Aerial apparatus are available in different shapes, sizes, and configurations, and they have various capabilities as well. This chapter is not intended to highlight all of the differences between the types of aerial apparatus currently on the market. Instead, considerations for utilizing these machines during master stream operations and options for maximizing their flow capabilities will be discussed. It is the responsibility of the readers to be familiar with the specific brand, configuration, and capacities of the aerial apparatus used by their fire department.

Aerial ladders

Aerial ladders are arguably the simplest of all aerial devices. They are equipped with interconnected ladder sections that telescope and provide continuous egress for firefighters and civilians.[1] Most fire departments use aerial ladders to access upper stories, but they can also be used for elevated master stream operations (fig. 8–1). Depending on the needs of the organization, aerial ladders may be equipped with a prepiped telescoping waterway mounted underneath the aerial device itself. Many organizations today

Figure 8–1. Aerial ladders are common tools used for delivering elevated master streams.

specify what is known as a pinnable waterway, which allows the monitor to be locked into either the last fly section or the second-to-last section. What is important to remember is that *NFPA 1900* does not require that a waterway be provided on an aerial ladder. In fact, some fire departments today still purchase aerial ladders without prepiped waterways.

NFPA 1900 requires that aerial ladders have a minimum *dry tip load* capacity of at least 250 lb.[2] While this is the minimum tip load allowed for aerial ladders, most manufacturers today produce units with tip loads of 500 lb or greater. Most manufacturers require that the tip load be reduced during elevated master stream operations. For aerial ladders that have prepiped waterways, it is equally important for the operator to know the maximum flow rate for the waterway. During elevated master stream operations, pump operators must ensure that they do not exceed the flow or pressure ratings of the waterway. Doing so could result in excessive nozzle reactions that could cause the aerial device to become overloaded. Most aerial ladders purchased by fire departments today come equipped with a waterway capable of flowing 1,000 gpm, although some manufacturers offer waterways capable of flowing 1,500 gpm.

This average maximum flow rate of 1,000 gpm for aerial ladders is comparable to the maximum flow rate of the deck gun found on most pumpers. For this reason, the operator should consider the monitor at the end of an aerial ladder as nothing more than an elevated deck gun. These monitors can be repositioned in various directions from either the ladder tip or the turntable pedestal while flowing. Depending on the age of the aerial ladder and what the department specified when it was built, the furthest up the monitor may be able to be pointed is parallel with the centerline of the aerial device. An apparatus configured this way makes it difficult to perform storefront blitz operations (fig. 8–2). They are usually positioned above the fire with their monitors pointed down during master stream operations.

Water towers

Water towers are equipped with a hydraulically operated boom that is primarily designed for elevated master stream operations. The vertical height of these telescoping or articulating aerial devices varies depending on the manufacturer, but units between 55' and 75' are common. *NFPA 1900* requires these pieces of equipment to have a permanently mounted waterway capable of delivering at least 1,000 gpm.[3] In contrast to other types of aerial apparatus, one of the advantages of water towers is that they are usually an

Figure 8–2. Some aerial ladders are specified to have their monitor sweep 30° above the ladder centerline.

option found on an engine company chassis. This allows the apparatus the ability to rapidly deploy an elevated master stream while still maintaining an overall length similar to a traditional pumper. Although some fire departments use water towers as pseudo truck companies, they are best suited to rapidly place an elevated master stream in service (fig. 8–3). Another major advantage of most telescoping water towers is their ability to position the boom low and direct the monitor upward for storefront firefighting operations.

Elevating platforms

Elevating platforms should be considered heavy artillery in the municipal firefighting world. These aerial apparatus are capable of performing a multitude of tasks on the fireground but are especially useful in providing heavy-caliber master streams. Depending on the needs of the fire department, elevating platforms can be designed as either telescoping, articulating, or combination aerial devices. Many organizations refer to elevating platforms as *tower ladders*, although this term describes elevating platforms that are specifically of *box boom construction* (fig. 8–4). Moving forward, the phrase *elevating platform* will be used to describe any type of aerial device that has a bucket attached at the end of the boom and is capable of providing heavy-caliber master streams.

Figure 8–3. Water towers can rapidly put an elevated master stream in service.

Figure 8–4. Tower ladders have a box-boom-constructed aerial device.

Regardless of the type of elevating platform used by an organization, all configurations must meet the same minimum standards outlined in *NFPA 1900.* With regard to master stream operations, operators must be mindful of a few specific requirements. Elevating platforms are required to have a minimum dry tip load rating of at least 750 lb. In addition, elevating platforms are required to have a prepiped waterway that supplies at least one monitor at the end of the bucket capable of flowing a minimum of 1,000 gpm. The standard also requires that these aerial devices be capable of maintaining a *wet tip load* of 500 lb while flowing 1,000 gpm from the bucket. Of particular importance for firefighting operations is that the standard requires a monitor in the bucket to be capable of moving at least 45° in both directions from center, as well as a minimum of 45° above and below the horizontal plane.[4]

Firefighters should be familiar with the difference between dry and wet tip loads. The dry tip load of any aerial device refers to the maximum amount of weight that can be placed at the tip of the aerial device when it is at 0° of elevation and is fully extended perpendicular from the chassis, with no water flowing from the aerial device itself. Wet tip load refers to the maximum amount of weight that can be placed at the aerial tip while the device is flowing water. The wet tip load is also determined with the aerial device at 0° of elevation while fully extended perpendicular to the vehicle chassis.

The operator must remember that the preceding information is considered to be the minimum requirements for elevating platforms. Many of the apparatus manufactured today are capable of dry tip load ratings of 1,000 lb and have waterways capable of supplying up to 2,000 gpm from the bucket (fig. 8–5). In addition, some elevating platforms purchased today come equipped with two permanently fixed monitors in the bucket. Considerations

Figure 8–5. Elevating platforms must be capable of maintaining a tip load of 500 lb during elevated stream operations.

for this type of bucket arrangement will be discussed in more detail later in this chapter.

Elevating platforms are formidable offensive tactical tools capable of knocking down tremendous fire loads. Since modern elevating platforms are capable of flowing massive volumes of water, feeding these beasts can be difficult from the standpoint of the supply pumper. While the tactics discussed in the remainder of this chapter can be applied to all types of aerial apparatus, the focus will primarily be on how to efficiently supply elevating platforms. Elevating platforms that are configured efficiently will be capable of incrementally increasing their flow rate based on the fire load present.

Quints

The majority of aerial apparatus purchased and manufactured today are considered *quint fire apparatus*. This term means that these types of apparatus are capable of performing five fireground functions and are equipped with a fire pump, a booster tank, an aerial device, ground ladders, and both supply and attack hose. Quints can be equipped with any of the aerial devices discussed above. From a water supply perspective, the major advantage of a quint is that its pump can be used for supplying water to one or more monitors at the end of the aerial device.

The intent of this text is not to discuss the pros and cons of quint fire apparatus. It is sufficient to understand that quints are highly popular in the U.S. fire service and can be particularly useful for master stream operations. Options for utilizing quint fire apparatus on the fireground during master stream operations will be discussed throughout this chapter.

Options for Supplying Aerial Apparatus

When it comes to supplying aerial apparatus for master stream operations, efficiency is paramount. Failure to properly supply the correct volume of water to an aerial apparatus will result in a lackluster stream that is incapable of overpowering the fire load. This wastes time and resources, while also placing firefighters and civilians at risk. Supplying water to aerial apparatus can essentially be broken down into three tactics. In this section, we will discuss each of these tactics and review how they can be effectively utilized on the fireground.

Direct aerial inlets

Many aerial apparatus are equipped with a *direct aerial inlet* that supplies water directly to the monitor at the tip of the aerial. On rear-mounted aerial apparatus, these direct inlets are typically located at the rear of the apparatus. Mid-mounted aerial apparatus, including *tillers*, or tractor-drawn aerials, have a direct aerial inlet located on the side of the apparatus, usually below the turntable (fig. 8–6). When an aerial apparatus is supplied through its direct aerial inlet, the pumper supplying the aerial is in complete control of the flow of water through the apparatus.

Using a direct aerial inlet is the only option for supplying ladder companies built with a prepiped waterway but no fire pump. If the aerial apparatus is a quint, the direct aerial inlet completely bypasses the pump. Although most quints are equipped with a direct aerial inlet, this is not the only way such apparatus are able to supply the waterway. Quints are required to have a discharge valve installed at the pump to control the flow of water to the aerial waterway.[5] This means that when supplying a quint, the operator of the truck company has the option of either connecting the supply line to the direct aerial inlet or to a pump intake.

When using a direct aerial inlet, the engine supplying the truck company is required to do all the work in order to produce an effective stream. This requires the pump operator to calculate the pump discharge pressure (PDP)

Figure 8–6. Direct aerial inlets are the only option for supplying aerial devices that do not have onboard pumps.

to appropriately account for the nozzle pressure of the monitor, the friction loss in the supply line, the appliance loss in the aerial, and the elevation loss between the monitor and the ground. When the supply engine and the aerial apparatus are close together and the flow rate is low, this is usually not an issue. If the supply engine is attempting to flow the maximum flow rate of the aerial device through several hundred feet of hose, however, the pump pressures required may be excessively high.

If the water supply is distant from the aerial apparatus, the supply operation will require two pumpers to operate efficiently. One pumper will be located at the water source and will supply water to the second pumper located near the aerial apparatus (fig. 8–7). The second pumper is responsible for pumping water through the direct inlet of the aerial apparatus at the appropriate pump pressure. In this scenario, the aerial device is nothing more than a giant monitor that the second engine is pumping through. Although this deployment of resources efficiently supplies water to the aerial device, three apparatus are required to achieve it—two pumpers and the aerial.

Due to the massive flow rates that elevating platforms are capable of achieving, it is not recommended to supply these monsters through the direct aerial inlet. This is primarily because a single engine that is any distance from the apparatus will struggle to produce the necessary pressure to effectively supply the maximum volume these machines are capable of delivering. If the elevating platform is a quint, the supply line should be connected into a pump intake on the rig. When dealing with an elevating

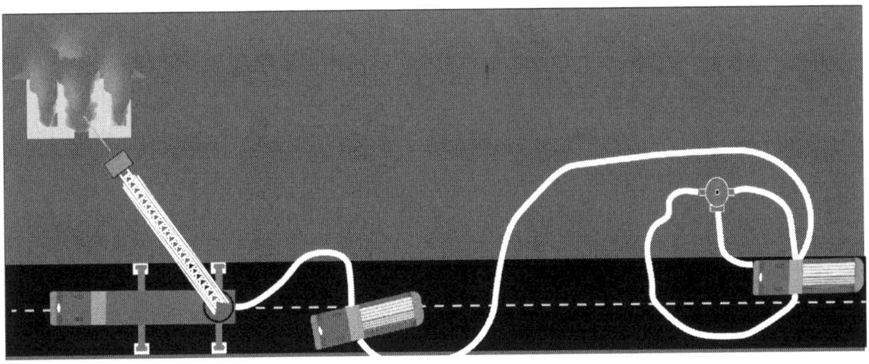

Figure 8–7. Aerial apparatus that are not equipped with a pump should be assigned two pumpers during water tower operations—one pumping the aerial device and the other supplying water from the source.

platform that is not equipped with a pump, it is critically important to assign two pumpers to the supply operation for maximum efficiency.

Aerial apparatus supplying itself

This option is only pertinent for quint apparatus and is typically performed in situations where the quint is the first apparatus to arrive on scene and elects to perform a forward lay from a hydrant. Some departments that rely on this tactic may connect the supply line directly to the hydrant and supply the quint with hydrant pressure alone. While this may be sufficient for handline operations off of the quint, it may prevent the full capacity of the aerial device from being achieved due to the friction loss in the supply line, especially when attempting to maximize an elevating platform.

This tactic can be made more efficient by connecting a four-way hydrant valve to the hydrant prior to supplying water to the quint (fig. 8–8). This will allow a later-arriving engine to pump the hydrant if the quint requires more volume. In the rare instance that a quint is extremely close to its own hydrant, it is acceptable to allow the apparatus to supply itself. This should

Figure 8–8. The four-way valve preconnected to the supply hose allows this quint to initially perform a forward lay on large-scale fires and be supplemented later in the incident.

only be done if the hydrant is within 100' of the quint. Otherwise, friction loss will reduce the volume available to the pump. The operator should employ a heavy hydrant hookup to ensure that the full rated capacity of the hydrant can be delivered into the apparatus.

Supply engine feeding the pump of an aerial apparatus

Utilizing a supply engine to feed the pump of an aerial apparatus is by far the most efficient option when supplying water to an aerial apparatus. It requires the aerial apparatus to be equipped with a pump, but it does not require the apparatus to be considered a quint. Some fire departments have built aerial apparatus that are equipped with a fire pump but carry a minimal amount of hose and have no booster tank. This configuration of an aerial apparatus may be chosen for a variety of reasons, but whatever the reason, it is a great option for elevating platforms. The primary purpose of the fire pump on the apparatus configured this way is to aid in supplying the full volume of water through the aerial device.

The most basic deployment of this type of supply operation has the engine company positioned at the water source with a supply line laid between the pumper and the aerial apparatus. The engine company will serve as the supply pumper and will pump water to the pump intake of the aerial apparatus. This creates a simple relay pumping operation between the supply engine and the aerial apparatus. The supply pumper only needs to focus on pumping the appropriate pump pressure to deliver the desired flow rate to the aerial apparatus and provide a sufficient intake pressure. The aerial apparatus will use its pump to build the required pressure to deliver the desired volume of water out of the monitor at the tip. When an engine is assigned to supply water to an aerial apparatus, it should indicate to the operator that large volumes of water are required. This should prompt the operator to employ the water supply tactics discussed in previous chapters to maximize the available water from the source.

Using an engine to pump water to the aerial's pump not only supplies water in the most efficient manner, but it also allows other types of supply operations to occur. In the event that another apparatus needs to put a master stream in service at the fire scene, a dual pumping operation can be implemented. In this scenario, the supply pumper will simply calculate the friction loss for the total flow rate at the scene and pump at the appropriate pressure to deliver it (fig. 8–9). Because intake-to-intake operations

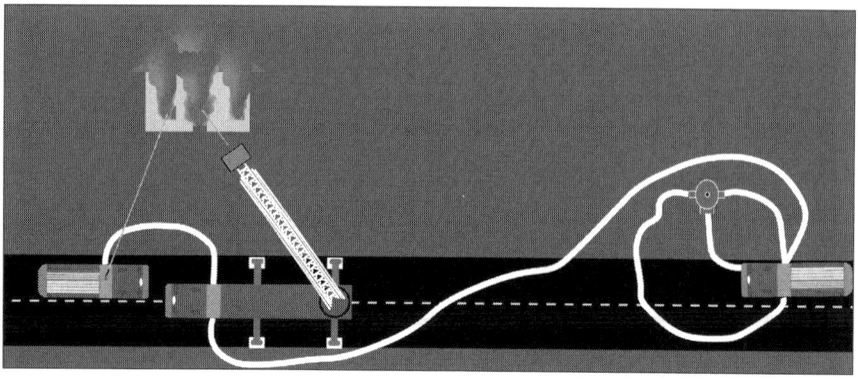

Figure 8–9. Using the pump of the quint to supply this operation allows a dual pumping operation to be performed by an additional pumper.

are self-regulating, the water will automatically divide between the two apparatus based on their flow demands.

Flowing the maximum volume of water that most modern elevating platforms are capable of providing requires extremely high pump discharge pressures. This is due to the friction loss within the aerial device at these flow rates and will be discussed in greater detail later in this chapter. By using a supply pumper to feed the pump of the elevating platform, however, a tandem pumping operation can easily be performed. This will help the aerial apparatus to build the pump pressures required to deliver the full flow capacity of the apparatus.

Fires in commercial and industrial buildings have a higher probability of requiring the use of elevated master streams. Operators of quint apparatus should anticipate this need and ensure that they use the most effective water supply tactics to achieve the highest possible flow rate. Among some in the U.S. fire service, there is a stigma against the use of quint fire apparatus, and organizations that run them often refuse to use them to their full potential. The author is not suggesting that they are the be-all and end-all solution for the fire service. They are not. What quints do provide is an extremely efficient and effective means of deploying heavy-caliber streams on the fireground. This is primarily because their pumps can be used to help build the pressures required for delivering large volumes of water from the aerial device. Rather than focusing on using quints for interior fire attack operations, operators should recognize how these types of apparatus can be effectively used for their ability to rapidly deploy heavy-caliber streams.

Flowing Water from the Aerial Device

Choosing the proper water supply tactic is paramount to the success of any elevated master stream operation. As discussed above, the ideal configuration involves assigning a dedicated supply engine to pump water from the source to the aerial apparatus. The water supply tactics utilized by the supply engine at the water source do not differ from those discussed in previous chapters. The only variable that may change between incidents is the type of water source from which the engine is operating. Nonetheless, there are a multitude of variables that the operator of the aerial apparatus must assess when performing an elevated master stream operation. This section will discuss these variables and will offer insight on how to effectively utilize modern aerial apparatus for elevated stream operations. The operator must understand that these considerations will differ based on the type of aerial apparatus being used, the age of the aerial apparatus, and the options found on that particular device.

Pre-1991 aerials versus modern aerials

Prior to 1991, there were no industry standards that applied to aerial apparatus. As a result, the American fire service witnessed a multitude of aerial failures. In an effort to prevent these catastrophic failures, minimum standards relating to the design, construction techniques, minimum tip load, and stability of aerial apparatus were implemented by the National Fire Protection Association. The current standards relating to aerial apparatus are delineated by *NFPA 1900*.[6]

Before we discuss tactics for elevated master streams, it is critically important for the reader to recognize that the tactics discussed are intended for modern aerial apparatus *only*. Operating pre-1991 aerial apparatus in the fashion outlined in the remainder of this chapter has a high potential of causing an aerial failure. The biggest difference between the two eras of aerial apparatus are the conditions in which they are designed to operate. Pre-1991 aerial apparatus were designed to operate on the vertical plane. With regard to elevated master stream operations, firefighters were taught to adhere to the *75/80/80 rule*. This guideline for master stream operations stated that the aerial device should be raised to an angle of 75°, extended no more than 80% of its full extended length, and have a smoothbore nozzle pumped at a nozzle pressure of 80 psi.[7] Failure to operate within these parameters could contribute to the collapse of the aerial device.

Aerial apparatus manufactured since 1991 are capable of being operated in a completely different fashion. From the standpoint of master stream operations, the reader must understand that modern aerial devices are designed and intended to flow water on the horizontal plane. Operating horizontally and directing water up and into the building on fire is the most efficient option for controlling large-scale fires. This is especially true when utilizing an elevating platform and is the mindset that will be used for elevated streams moving forward (fig. 8–10). Prior to using an aerial apparatus in this configuration, the reader must first determine if the aerial apparatus in question is safely capable of doing so.

Nozzle selection

The nozzle utilized at the end of an aerial device has a profound impact on the efficiency of a heavy-caliber firefight. There are two choices fire departments have regarding the type of nozzle utilized on the aerial—smoothbore or combination fog nozzles. In the fire service, a great debate revolves around the use of these two nozzle types, especially with regard to interior firefighting. In terms of master stream operations and aerial devices, however, smoothbore nozzles are the best option for aggressively fighting large bodies of fire.

There are instances where it is beneficial to have a combination fog nozzle at the end of the aerial device. These can be helpful during exposure

Figure 8–10. Modern aerial apparatus are designed to operate on the horizontal plane, and this is particularly true of elevating platforms.

protection or for mass decontamination at a hazardous materials incident. During a hazmat incident there is usually time for the aerial operator to switch out the smoothbore nozzle and replace it with a fog nozzle for mass decontamination. While aerial apparatus can certainly be used for exposure protection, incident commanders should ask themselves an important question: "Do I want to use my one or two million-dollar aerial apparatus to do the job that a ground monitor can do, or should I use this limited resource to punch the fire in the face?"

The major advantage to using a smoothbore nozzle for elevated stream operations is the fact that these streams provide more reach and penetration at lower nozzle pressures (fig. 8–11). This allows the firefighters operating these nozzles to place the stream more effectively. In addition, because the stream exiting a smoothbore nozzle is solid and compact, more of the water is capable of surviving the high temperatures near the ceiling of the container and thus reaching the seat of the fire. The major disadvantage of smoothbore nozzles is that increasing the flow rate from the aerial device usually requires a tip to be removed. While it is possible to pump smoothbore nozzles at a higher nozzle pressure to gain some additional volume, this practice comes with its own set of concerns and will be discussed in another section. Due to their effectiveness during elevated stream operations, smoothbore master stream nozzles will be referenced for the remainder of this chapter.

Figure 8–11. Smoothbore nozzles are the nozzles of choice for elevated master stream operations.

Operators must be familiar with the nozzle pressure and flow rates for the smoothbore tips used on their aerial apparatus. The general guideline is to pump smoothbore master streams at a nozzle pressure of 80 psi. Table 8–1 lists the common master stream tips found on modern aerial ladders and elevating platforms, as well as their flow rates. Recall that the waterways and monitors of most aerial ladders are capable of flowing the same volume of water as a pumper's deck gun. Using this analogy, a good starting tip for an aerial ladder is the 1⅜" tip. At 80 psi, this will deliver a starting flow rate of 500 gpm. Most smoothbore quad stacks have a 1⅜" tip as the starting tip and a 2" tip as the base tip. Usually in between these two tips are the 1½" and 1¾" tips. Referring to table 8–1, we see that switching from a 1⅜" tip to a 1½" tip only gives us roughly an additional 100 gpm, rendering this switch arguably useless.

For ease of operations, the author recommends only using two tips on the quad stack set: the 1⅜" tip and the 2" tip. Operations should begin with the 1⅜" tip, which will allow for 500 gpm to easily be placed in service. If the needed fire flow at the scene must be increased, the water supply can be built out to accommodate the flow requirements. Once this is done, the aerial ladder can simply switch the 1⅜" tip for the 2" tip to increase the flow. From a simplicity standpoint, the major advantage of this is that an exact doubling of the flow rate occurs while maintaining the same nozzle pressure. The mindset should be to "double the flow until it dies." Thus if 500 gpm is not working, the flow rate should be increased to 1,000 gpm.

Elevating platforms are capable of flowing greater quantities of water, and the stack sets found on these aerial devices reflect this capability. The most common tower ladder stack set consists of four tips: 2", 2¼", 2½", and 2¾" (fig. 8–12). This stack set is capable of a starting flow of 1,000 gpm through the 2" tip and approximately 2,000 gpm through the 2¾" tip at

Table 8–1. Common smoothbore master stream tip flow rates

Tip size (in.)	Nozzle pressure (psi)	Flow rate (gpm)
1⅜	80	502
1½	80	597
1¾	80	813
2	80	1,062
2¼	80	1,345
2½	80	1,660
2¾	80	2,010

Figure 8–12. A common stack set found on many elevating platforms today. The starting tip is 2" in diameter, and the base tip is 2¾".

80 psi nozzle pressure. Although this configuration is certainly capable of flowing the rated capacity of most elevating platforms, it can create some problems.

One of the issues that occurs when elevating platforms start their stack sets with a 2" tip is that it requires at least 1,000 gpm to be available from the water source in order for an effective stream to be produced. Stream quality and reach will suffer tremendously if the water supply operation is unable to support this flow rate initially. In the urban setting, it is possible for the first hydrant utilized for the operation to have a low flow capacity. These situations will require that an additional water source be established in order to supplement the flow. Other situations may occur where the hydrant is capable of providing the necessary volume, but there is too much friction in the supply line between the hydrant and the elevating platform. This is why it is so important to assign a supply pumper to the hydrant or utilize a four-way hydrant valve. In the rural setting, it is unlikely that a water shuttle operation will be capable of immediately supporting a 1,000 gpm flow rate from the elevating platform. The result in all of these situations will be pitiful streams that do not reach the seat of the fire and will simply become wastewater.

At the opposite end of the spectrum is the use of the 2¾" tip. While this tip is able to single-handedly flow the rated waterway capacity of most elevating platforms currently manufactured, it comes at a cost. The biggest issue that stems from using the 2¾" tip is the horrendous stream that is usually produced, which is often referred to as water vomit. The stream leaving the nozzle has no shape and resembles a narrow fog pattern rather than a solid stream (fig. 8–13). In order to understand why this occurs, we must first evaluate the monitors used on most elevating platforms. Monitors capable of flowing 2,000 gpm that are installed on elevating platforms typically have an outlet opening measuring 3½" in diameter. This outlet is equipped with threads to allow the nozzle of choice to be attached.

When an automatic fog nozzle is attached to these 2,000 gpm monitors, no real issue can be detected. By design, an automatic fog nozzle is able to maintain its pattern by varying the flow out of the nozzle depending on the pressure entering the nozzle base. The baffle within an automatic nozzle is designed to break up the stream into smaller droplets of water before it exits the nozzle. As a result, the streams leaving these nozzles are always broken up to some degree even on the straight stream setting. The intention behind use of a smoothbore nozzle for master stream operations is to take advantage of its ability to maintain a tight, solid pattern that is able to penetrate superheated environments and reach the seat of the fire. This

Figure 8–13. Water vomit from a 2¾" smoothbore tip

does not occur when a 2¾" tip is used on a monitor with a 3½" outlet opening because an insufficient amount of shaping occurs within the nozzle to prevent the stream from breaking apart almost immediately after exiting the tip.

The generally accepted guideline for producing an effective stream from a smoothbore nozzle is that the tip diameter should not exceed more than one-half the diameter of the hose or pipe that is supplying it.[8] The whole purpose of a nozzle is to shape the stream into something usable for firefighting operations. When the nozzle opening is close in size to the hose or pipe supplying the tip, little to no shaping of the stream occurs, and so-called water vomit is produced. In addition, if the length of the tip is relatively short, this problem is further exacerbated. This is exactly what happens when the 2¾" tip is used on the majority of monitors found on elevating platforms today.

Technically speaking, the largest tip that should be on monitors with a 3½" outlet is a 1¾" tip, but it is acceptable to consider the 2" tip as the largest tip to use on a single monitor. When comparing the outlet size of a 2" tip with a 3½" monitor, we find that the 2" tip is 57% of the total outlet area of the monitor feeding the nozzle. Exceeding the recommended area by 7% is much better than the results obtained with a 2¾" tip, which exceeds it by roughly 28.5%. By considering the 2" tip the largest tip to use on these monitors, the elevating platform is capable of producing a compact fire stream that is easily able to penetrate a superheated environment and extinguish large volumes of fire. In addition, it makes flow calculations quite easy since the 2" tip flows 1,000 gpm at a nozzle pressure of 80 psi.

When configuring the stack set of an elevating platform, care and planning are crucial. Given the factors previously discussed, it is recommended that the starting tip on an elevating platform be the 1⅜" tip. Just as with an aerial ladder, it allows a compact, highly focused 500 gpm stream to be rapidly placed in service (fig. 8–14). If the flow rate must be increased, the firefighter in the bucket can simply switch out to the 2" tip. This will effectively double the flow rate from the elevating platform without sacrificing the efficiency of the stream (fig. 8–15).

Fire departments should be mindful of just adding a quad stack on top of the tower ladder stack that may already be present on the apparatus. Doing so may result in an extremely long nozzle tip that could potentially block the access doors of the platform. The better option is to attach a 3½" to 2½" reducer with an integrated stream shaper to the monitor outlet. Then

Figure 8–14. Starting off with a 1⅜" tip in the bucket of an elevating platform allows a laser-accurate stream to be placed in service while the water supply is expanded.

Figure 8–15. Switching to the 2" tip allows twice the volume of water to be flowed from the bucket without sacrificing functionality.

the quad stack with the 1⅜", 1½", 1¾", and 2" tip can be attached. This will allow the monitor to maintain a smaller profile while also enabling it to support the desired flow rates.

There may be instances in which a department wants to flow 2,000 gpm out of their elevating platform. Although most modern elevating platforms are equipped with waterways and monitors capable of flowing 2,000 gpm, doing so results in an almost unusable stream. In order to achieve a 2,000 gpm flow rate from an elevating platform efficiently, dual platform monitors are recommended.

Dual monitors on elevating platforms

Elevating platforms equipped with two heavy-caliber monitors are a common sight in the U.S. fire service. Few organizations utilize this configuration to its full potential, however. It is not uncommon to see dual monitors of elevating platforms equipped with one automatic fog nozzle and one smoothbore nozzle. The thought process is that the elevating platform can place either monitor in service based on the needs of the incident. While there is some merit to this, it is not the most effective use of a dual monitor configuration.

The true purpose of two monitors in the bucket of an elevating platform is the ability to efficiently flow large volumes of water using both monitors. When two different types of nozzles are utilized to do this, several problems occur. The first is due to the pressure differential that occurs. Smoothbore master streams operate at a nozzle pressure of 80 psi, while most automatics operate at 100 psi. This will result in overpumping one of the nozzles and underpumping the other. Since most departments that operate this way use an automatic nozzle on one of the monitors, the impact is not as severe. The automatic nozzle will still have an acceptable pattern, although the flow rate may not be the same as with the smoothbore nozzle.

It should be noted that some manufacturers produce automatic adjustable pressure nozzles that allow the nozzle pressure to be changed based on the needs of the incident. For example, the operator could set the nozzle pressure to a higher setting to allow a lower volume of water to flow a further distance, or the nozzle pressure could be set lower to match that of a smoothbore nozzle. Although this allows the nozzle pressures of both monitors to match, it will not change the fact that the two streams produced are quite different from each other. For these reasons, it is recommended that elevating platforms equipped with two monitors outfit them both with smoothbore nozzles and the tip sizes previously mentioned (fig. 8–16). This provides the tower company with several benefits for operations with heavy-caliber streams.

Master stream operations typically begin with only one of these monitors flowing. Since the recommended starting tip on the monitor is a 1⅜"

Figure 8–16. Elevating platforms with dual monitors should be configured with smoothbore nozzles for each monitor.

tip, the initial flow rate will be 500 gpm. If the elevating platform has to increase the flow rate to 1,000 gpm, it can be done one of two ways. Option one is to shut down the first monitor and switch to the 2" tip. This is a great option if there is only one firefighter in the bucket to operate the monitor. The other option is to simply open up the second monitor that is also equipped with a 1⅜" tip. Both choices result in the flow rate doubling from the bucket, but opening the second monitor provides the ability to direct an additional stream elsewhere.

In order to efficiently flow 2,000 gpm from an elevating platform, the use of both monitors should be strongly considered. Both monitors should be equipped with 2" tips, which will allow 1,000 gpm to flow from each monitor. Not only will this permit the full flow capacity of the apparatus to be delivered from the bucket, but the streams produced from each monitor will be more compact and usable for the firefighting operation (fig. 8–17).

When building or purchasing an elevated platform, fire department personnel often wonder if two monitors are necessary. In order to make this determination, they must first evaluate their target hazards and their water supply capabilities. Since many apparatus manufacturers provide a 2,000 gpm waterway as a standard option on their elevating platforms, the fire department must ask whether or not they have target hazards that could require this type of flow rate. If the answer is yes, then department

Figure 8–17. This elevating platform is flowing 2,000 gpm from a bucket using dual monitors. Each monitor is equipped with a 2" tip.

personnel should consider dual monitors in order to provide this flow safely and efficiently from the aerial apparatus. If they determine that they will not need to flow much more than 1,000 gpm or that their water supply is incapable of reliably providing a 2,000 gpm flow, then a single monitor in the bucket may be sufficient. The key here is to remember that dual monitors allow these monsters to flow their rated capacity while producing usable, highly accurate firefighting streams.

Load charts and water tower operations

Understanding the load capabilities of an aerial device is a critical responsibility of the aerial operator. As discussed earlier, *NFPA 1900* outlines the minimum dry tip load requirements that all types of aerial devices must be capable of meeting. In addition, the standard requires that elevating platforms be capable of maintaining a tip load of at least 500 lb while flowing water from the bucket. The standard also requires the monitor to be capable of flowing at least 1,000 gpm while it is aimed at the following angles, depending on the aerial type:[9]

- *Aerial ladders equipped with a waterway*
    - Below aerial device centerline—135°
    - Left and right of aerial device centerline—at least 45°
- *Water towers*
    - Above aerial device centerline—at least 30°
    - Below aerial device centerline—135°
    - Left and right of aerial device centerline—at least 45°
- *Elevating platforms*
    - Above aerial device centerline—at least 45°
    - Below aerial device centerline—at least 45°
    - Left and right of aerial device centerline—at least 45°

Aerial operators must remember that any time a monitor is flowing water, there is a reaction force in the opposite direction of the monitor. The reaction force applied to the aerial device is dependent on the volume of water exiting the nozzle, the nozzle pressure, and the angle at which the monitor is pointed. While the values listed above are considered the minimum requirements for different types of aerial apparatus, some manufacturers have designed their aerials to withstand reaction forces at wider angles. Nevertheless, wider angles should not be used unless the apparatus manufacturer specifically says that it is safe to do so.

Aerial operators must refer to the load chart on their particular aerial apparatus to determine the acceptable monitor angles and flow rates. For example, some modern aerial ladders are capable of raising their monitor 30° above the centerline of the aerial device for storefront firefighting operations. Doing so may put the tip load at zero pounds due to the reaction force applied to the aerial ladder, however. If your aerial ladder is not capable of doing this, *do not* try to make it do so by placing a 30° elbow on the end of the monitor. This could subject the aerial ladder to a reaction force that it was not designed to withstand and could result in a catastrophic aerial failure.

Certain manufacturers of elevating platforms have no restrictions on the position of the monitor as long as the flow rate remains below a certain threshold. Once the flow rate exceeds this threshold, certain monitor angles are restricted (fig. 8–18). One manufacturer in particular lists this threshold as 1,500 gpm. At flow rates less than 1,500 gpm, the monitors can be pointed in any direction. Flow rates greater than 1,500 gpm require that the monitors do not operate outside of the specified operating angles. Failure to adhere to this could cause the elevating platform to become overloaded.

When dealing with dual platform monitors, the aerial operator must understand how the position of each monitor impacts the aerial device. The flow rate through each of the monitors should be determined first. If each monitor has a 1⅜" tip and is flowing, then the total flow rate from the bucket would be 1,000 gpm. Assuming we are operating an elevating platform with

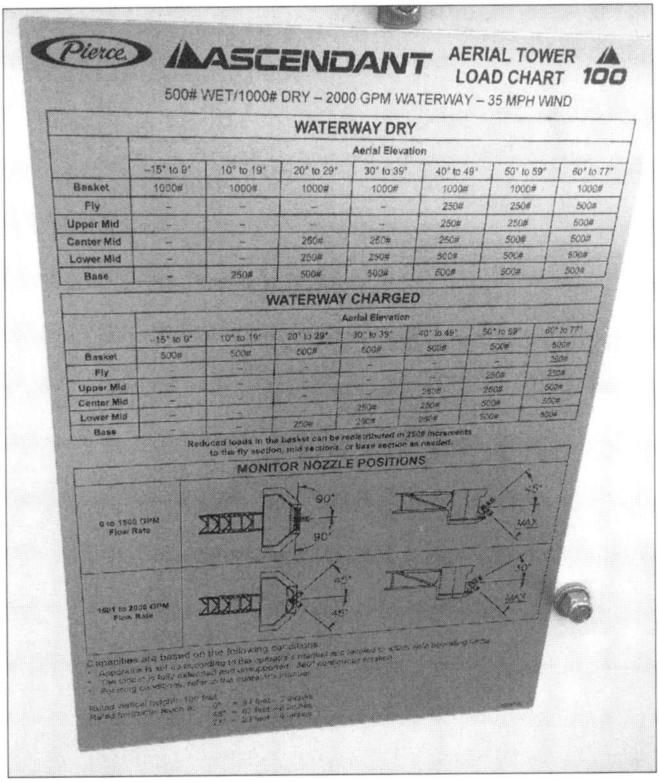

Figure 8–18. The load chart on this elevating platform specifies monitor angles at certain flow rates.

a 1,500 gpm threshold limit, these two monitors can be pointed in any direction required. Placing 2" tips on both monitors would deliver a 2,000 gpm flow rate from the bucket. Since this is above the threshold limit, however, the firefighters in the bucket must be mindful of monitor positioning.

For example, the load chart may specify that flows greater than 1,500 gpm cannot be directed more than 45° left or right of the aerial centerline. This would mean that both monitors could not be pointed more than 45° to the right of the bucket. Nevertheless, it would be acceptable to direct one monitor 45° up and the other 90° to the right. This would cause 1,000 gpm to be directed 45° upward, with another 1,000 gpm aimed 90° to the right of the bucket. As long as no more than 1,500 gpm is directed outside of the threshold zone, then there is no issue with this operation. Operating in this fashion ultimately allows for a larger operating envelope while flowing large volumes of water from the bucket (fig. 8–19).

Overpumping smoothbore nozzles

Currently in the U.S. fire service, there is a discussion regarding the overpumping of smoothbore master stream nozzles. This is referred to as *high velocity stream operations*, and it simply means pumping these nozzles at a nozzle pressure greater than 80 psi. Increasing the nozzle pressure of a smoothbore nozzle also increases the following:

Figure 8–19. The major advantage of dual platform monitors is the ability to point heavy-caliber streams in multiple directions.

- Stream velocity
- Stream reach
- Nozzle reaction
- Flow rate

The major advantage to overpumping smoothbore nozzles is that it increases the stream velocity and stream reach, while only minimally increasing the flow rate from the nozzle tip. The theory behind this tactic is that by increasing stream velocity, more water is able to reach the seat of the fire because a smaller percentage of it evaporates as it travels through the superheated environment. It should be noted that this concept is currently only a theory, and its usefulness and effectiveness has yet to be verified through actual scientific field testing. It is being discussed here to highlight the impact that it has on the reaction forces applied to aerial apparatus.

Table 8–2 summarizes how doubling the nozzle pressure of a smoothbore master stream affects the flow rate, nozzle reaction, and stream velocity. The information contained in this table shows that doubling the nozzle

Table 8–2. Effects of increased nozzle pressure on flow rate and nozzle reaction of smoothbore master streams

Tip size (in.)	Nozzle pressure (psi)	Flow rate (gpm)	Nozzle reaction (lb)	Stream velocity (mph)
1⅜	80	502	237	74
	160	710	475	104
1½	80	598	283	74
	160	845	565	104
1¾	80	814	385	74
	160	1,151	769	104
2	80	1,062	502	74
	160	1,503	1,005	104
2¼	80	1,344	636	74
	160	1,902	1,272	104
2½	80	1,659	785	74
	160	2,348	1,570	104
2¾	80	2,008	950	74
	160	2,841	1,900	104

pressure results in a 40.5% increase in stream velocity and a 41.5% increase in the flow rate when the tip size remains constant. Of greater importance is the fact that when the nozzle pressure doubles, so does the nozzle reaction. This means that overpumping a smoothbore tip at the end of an aerial apparatus could easily exceed the tip rating and overload the ladder. This will be dependent on the tip size and how much the nozzle pressure is increased above the standard 80 psi applied for smoothbore master streams.

Operators must understand the impact that overpumping master stream nozzles has on the aerial device. In addition, it is critically important for the aerial operator to be highly knowledgeable concerning the information contained in the aerial device's load chart. The operator should remember the following if this tactic is to be utilized with an aerial device:

- It is vital to ensure that no reaction forces ever exceed the manufacturer's recommendations.
- At a nozzle pressure of 80 psi, the reaction force value is roughly one-half that of the flow rate (i.e., a 1⅜" tip flows 500 gpm at 80 psi, and the reaction force is 237 lb).
- At a nozzle pressure of 160 psi, the reaction force value is nearly the same as the value of the flow rate at 80 psi (i.e., a 1⅜" tip being pumped at 160 psi has nearly 500 lb of nozzle reaction).

In order to perform this technique safely, it is recommended that the tip size of the nozzle not exceed 1⅜", and the nozzle pressure not exceed 160 psi. Depending on the monitor utilized, the rated operating pressure may not be much more than 160 psi. Operating above this could cause the monitor to fail disastrously. When using aerial devices for high velocity stream operations, a tip size of 1⅜" should not be exceeded to ensure that the reaction forces placed on the aerial device do not exceed the aerial's capabilities. Recall that the NFPA standard requires aerial ladders equipped with a waterway to be capable of supplying at least 1,000 gpm.[10] Using a smoothbore nozzle traditionally, this would be accomplished with a 2" tip pumped at 80 psi nozzle pressure, with a resulting reaction force of 502 lb. If a 1⅜" tip is used instead and is pumped at a nozzle pressure of 160 psi, the flow rate would be 710 gpm, and the nozzle reaction would be roughly 475 lb. This would allow a stream with more exit velocity to leave the nozzle without exceeding the limits of the aerial device.

A 1⅜" tip should also be used when performing high velocity stream operations from an elevating platform. One advantage of doing so is consistency in the deployment of the operation regardless of the type of aerial device used. Using the 1⅜" tip on the monitor of an elevating platform also minimizes the likelihood of having to limit monitor positions and angles due to the reaction forces. Even when pumped at 160 psi, the reaction force from the 1⅜" tip will not exceed that of the 2" tip pumped at 80 psi. This will allow the firefighters in the bucket to direct the high velocity stream safely and effectively wherever it is required.

Appliance loss

When supplying an elevated master stream, the operator must ensure that the pump discharge pressure accounts for appliance loss through the aerial device. Historically, firefighters have been taught to apply 25 psi for the appliance loss when pumping an elevated stream. This general rule typically only applies to master stream devices flowing up to 1,000 gpm. According to *NFPA 1900*, the appliance loss in the waterway of an aerial apparatus should not exceed 100 psi while flowing 1,000 gpm.[11] Technically speaking, this means that the internal appliance loss of an aerial apparatus could range between 25 psi and 100 psi for flows up to 1,000 gpm. This highlights the importance of flow testing aerial devices to ensure they are pumped at the appropriate pressures.

We know that friction loss increases exponentially with an increase in the flow rate. Specifically, when the flow rate doubles, the friction loss through the conduit quadruples. This holds true whether the water is flowing through fire hose or through an aerial waterway. Many elevating platforms manufactured today are capable of flow rates in the 2,000 gpm range. Only adding 25 psi for appliance loss through the aerial device at a flow rate of 2,000 gpm will result in a monitor that is severely underpumped. This ultimately leads to an inefficient fire stream from the platform.

NFPA 1900 does not specify a maximum friction loss allowance for elevating platforms at 2,000 gpm flow rates. Nonetheless, internal appliance loss values between 100 psi and 120 psi are not unusual while flowing 2,000 gpm through modern tower ladders (fig. 8–20). This should again be verified through flow tests. An operator who knows the appliance loss of the elevating platform at a flow rate of 1,000 gpm can multiply that value by four to get a ballpark estimation of the appliance loss at 2,000 gpm. The

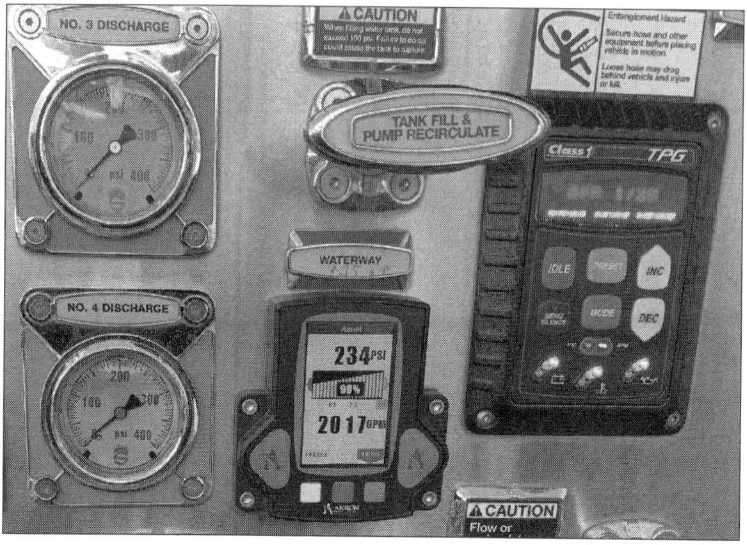

Figure 8–20. In order to flow 2,000 gpm from this tower ladder, a pump discharge pressure of 234 psi was required on the pump of the aerial. The aerial device was at 0° of elevation.

key here is to understand that it requires high pump discharge pressures in order to achieve 2,000 gpm flow rates out of most elevating platforms.

Another important consideration when flowing larger volumes from elevating platforms is the appliance loss within the monitor. The exact appliance loss values will vary based on the model and brand of monitor, which is why it is important to refer to the monitor manufacturer to obtain specific appliance loss values. Typically, monitors capable of flowing 2,000 gpm can flow 1,000 gpm with a negligible amount of loss. At a flow rate of 2,000 gpm, however, the appliance loss from the monitor alone can be 30 psi or greater depending on the monitor used, which is in addition to the waterway appliance loss.[12] Taking each of these variables into account, we can determine the pump discharge pressure required to flow 2,000 gpm out of a hypothetical elevating platform.

Example 8–1. You are operating a tower ladder that is being supplied by another pumper. The supply line is entering your pump intake, and you are tasked with supplying 2,000 gpm from a single monitor in the platform. The monitor has a 2¾" tip and approximately 30 psi of loss when flowing 2,000 gpm. The waterway of your aerial apparatus has roughly 120 psi of appliance loss at a flow rate of 2,000 gpm. The platform is positioned just above the ground for a storefront blitz attack. Determine the pump discharge pressure required to supply the aerial.

$$PDP = \textit{Nozzle Pressure}\ (NP) + \textit{Total Friction Loss}\ (TFL) + \textit{Appliance Loss}\ (AP) \pm \textit{Elevation Loss or Gain}\ (EL)$$

where

$$NP = 80 \text{ psi}$$
$$TFL = 0 \text{ psi}$$
$$\textit{Monitor } AP = 30 \text{ psi}$$
$$\textit{Waterway } AP = 120 \text{ psi}$$
$$EL = 0 \text{ psi}$$

Substituting these values in the preceding equation:

$PDP = 80 + 0 + 30 + 120 + 0$

$\boldsymbol{PDP = 230}$ **psi**

In example 8–1, we see that the pump discharge pressure required to supply 2,000 gpm out of a single monitor is approximately 230 psi. The engineer operating the pump panel of the quint would have to set the aerial discharge gauge at this pressure. The reason that no value was entered for total friction loss is because there is no hose used in this example. The water is flowing from the quint's pump, through the waterway, and to the monitor. This is also why there are two separate appliance loss values. One is for the monitor at 2,000 gpm, and the other is for the waterway at 2,000 gpm.

Example 8–2. You are operating a tower ladder that is being supplied by another pumper. The supply line is entering your pump intake, and you are tasked with supplying 2,000 gpm from two monitors in the platform. Each monitor is equipped with a 2" tip and has an appliance loss value of roughly 7 psi at 1,000 gpm. The waterway of your aerial apparatus has roughly 120 psi of appliance loss at a flow rate of 2,000 gpm. The platform is positioned just above the ground for a storefront blitz attack with both monitors flowing. Determine the pump discharge pressure required to supply the aerial.

$$PDP = NP + TFL + AP \pm EL$$

where

$$NP = 80 \text{ psi}$$
$$TFL = 0 \text{ psi}$$
$$\textit{Monitor } AP = 7 \text{ psi}$$
$$\textit{Waterway } AP = 120 \text{ psi}$$
$$EL = 0 \text{ psi}$$

Substituting these values in the preceding equation:

$PDP = 80 + 0 + 7 + 120 + 0$
$\boldsymbol{PDP = 207}$ **psi**

By flowing 1,000 gpm out of each monitor, the operator is able to supply the desired 2,000 gpm flow rate from the bucket and simultaneously reduce the appliance loss in each monitor. This effectively lowers the pump discharge pressure required for the operation. The reduction in pressure gives the aerial operator the flexibility to raise the aerial device, if necessary, without sacrificing the flow rate or stream quality. For example, elevating the bucket 40' in the air would place the needed pump discharge pressure at 230 psi, which is the same pressure required to flow 2,000 gpm out of one monitor. The use of dual monitors from the platform not only makes high volume stream placement more versatile, but it also makes supplying these large flow rates more realistic.

Due to the high pressures that must be developed to support a 2,000 gpm flow rate out of an elevating platform, a tandem pumping operation will likely be required. This will help to reduce the total amount of pressure that must be developed by the quint, thus reducing its engine revolutions per minute. In order to accomplish this, supply engines should pump at a higher discharge pressure to provide more residual intake pressure at the quint. This efficiently balances the workload between multiple apparatus.

Study Questions

1. Which of the following types of aerial apparatus typically have the highest flow capacity through the aerial device?

   a. Aerial ladders
   b. Water towers
   c. Elevating platforms
   d. Tractor-drawn aerials

2. Flowing water through a direct aerial inlet is required for which type of aerial apparatus?

   a. Quints
   b. Aerial apparatus with no pump

c. Elevating platforms with a pump, but no booster tank
d. All of the above

3. With regard to elevated master stream operations, modern aerial apparatus are designed to:

a. Flow water according to the 75/80/80 rule
b. Have the monitor remain parallel with the centerline of the aerial device
c. Flow water on the horizontal plane
d. Only flow 500 gpm

4. Which of the following is an advantage of dual monitors on an elevating platform?

a. Allows 4,000 gpm to flow from the bucket
b. Reduces the nozzle reaction exerted on the aerial device
c. Allows the total flow rate to be split into two useable streams
d. Increases the pump discharge pressure required to flow water from the bucket

5. According to *NFPA 1900*, elevating platforms are required to maintain a tip load of __________ while flowing water.

a. 250 lb
b. 500 lb
c. 750 lb
d. 1,000 lb

6. Which of the following is a disadvantage of overpumping a smoothbore master stream nozzle on an aerial device?

a. Increases the stream velocity
b. Increases the stream reach
c. Increases the flow rate
d. Increases the nozzle reaction

7. Which of the following master stream tips does the text recommend should be the largest tip used for high velocity stream operations?

a. 1⅜"
b. 2"
c. 2½"
d. 2¾"

8. Generally speaking, an appliance loss value of 25 psi is considered acceptable for flows up to _______ through an aerial device.

    a. 500 gpm
    b. 1,000 gpm
    c. 1,500 gpm
    d. 2,000 gpm

9. When flowing 2,000 gpm through an elevated platform, the appliance loss within the aerial device can be as high as _________.

    a. 25 psi
    b. 50 psi
    c. 120 psi
    d. 250 psi

10. An elevating platform is flowing 2,000 gpm from its bucket using two monitors. Each monitor is flowing 1,000 gpm. Which of the following statements is true regarding the impact of this configuration?

    a. It increases the nozzle reaction exerted on the bucket compared to flowing the total flow from a single monitor.
    b. It requires the pump operator to double the nozzle pressure to account for the second monitor.
    c. It has no impact on the operation.
    d. It reduces the appliance loss within the monitors, thereby reducing the necessary pump discharge pressure required for the operation.

9

LARGE FLOW OPERATIONS AND TACTICS

Fires in large commercial buildings, warehouses, industrial facilities, and crowded urban areas have the potential to need extremely high fire flows. While these incidents can be categorized as once-in-a-career fires, they will require all of the firefighters operating on the scene to be unequivocally proficient in their duties. Pump operators especially will be forced to put their skills to the test in order to efficiently support the flow demands of the incident.

This chapter will discuss the various water supply tactics that can be implemented during these large-scale incidents (fig. 9–1). All of these advanced water supply tactics require the operator to be proficient with

Figure 9–1. Large-scale fires will require operators to maximize any water source they can access.

the pumping techniques discussed in the previous chapters with regard to drafting and hydrant operations. Most importantly, these large flow evolutions need to be practiced by the fire department. Waiting until one of these incidents occurs is not the time to try to implement these water supply operations for the first time. By the conclusion of this chapter, the reader will have a firm understanding of how to work as part of these more complex water supply operations and will be aware of the resources required to implement them and the considerations for their success.

Parallel Supply Line Operations

With regard to large water delivery, we know that the most efficient way to increase the available water entering the pump is to increase the total area feeding the pump. During hydrant operations, this involves performing a heavy hydrant hookup, and in the rural environment, it requires a multiple intake drafting operation. When flowing large volumes of water on the discharge side of the pump, the same principles apply. With a larger discharge area, higher volumes of water can be efficiently discharged. This can be accomplished through a *parallel supply line operation* from a strong water source.

Parallel supply line operations allow higher flow rates to be delivered to the fire scene compared to the volume a single large diameter hose (LDH) supply line is capable of providing (fig. 9–2). These operations involve laying a secondary large diameter supply line between the water supply pumper and the attack engine. The key to the success of this tactic is that the water source must be capable of supplying the volume of water that the supply pumper is attempting to deliver. After determining that the source is capable of the desired volume, the pump operator must ensure that the appropriate intake connections are made to allow that volume into the pump.

Table 9–1 lists the minimum discharge flow rates based on the plumbing size of the discharge according to *NFPA 1900*.[1] The operator must remember that these are the minimum flow rates that discharges must be capable of supplying. Depending on how the apparatus is built and plumbed, higher flow rates from a particular discharge may be possible. This is why it is critically important for the fire department to take the time to flow test and determine the maximum realistic flow rate for each individual discharge on the apparatus.

Figure 9–2. This elevating platform is being supplied by a parallel supply line operation and is flowing 2,000 gpm.

Table 9–1. NFPA 1900 discharge rates based on outlet size

Outlet size (in.)	Discharge flow rate (gpm)
2½	250
3	375
4	625
5	1,000
6	1,500

Source: Adapted from NFPA, *NFPA 1900: Standard for Aircraft Rescue and Firefighting Vehicles, Automotive Fire Apparatus, Wildland Fire Apparatus, and Automotive Ambulances, 2024,* p. 112 (Quincy, MA: NFPA, 2023).

Most municipal pumpers in service today are equipped with a single large diameter discharge with either 3" or 4" plumbing. According to *NFPA 1900*, a 3" discharge only needs to be capable of supplying 375 gpm, while a 4" discharge only needs to be capable of supplying 625 gpm.[2] It is usually possible to flow more than the minimum listed flow rates through large diameter discharges. The maximum realistic flow rate of most large diameter discharges is typically 1,000 gpm, however. Flows greater than 1,000 gpm cause an exponential increase in the internal appliance loss of the pumper, making these flow rates unrealistic in most cases. Even with the use of LDH, the limiting factor tends to be the discharge plumbing. For these reasons, it is recommended that firefighters consider the maximum realistic volume of a single supply line evolution to be 1,000 gpm. Fire scenes requiring more water from a particularly strong source will require the implementation of a parallel supply line operation.

Laying an additional supply line may seem like a daunting task, but it will be required if the fire scene requires the full flow capacity of a particularly strong hydrant. Depending on the distance between the water source and the fire scene, the second supply line may be able to be *hand jacked*, or manually laid out, between the two engines. Further distances may require a later-arriving engine to lay a supply line between the water source and the fire scene by either a forward or reverse layout. This will depend on the direction from which the third engine approaches. If units are tasked with establishing a secondary or tertiary water source from a strong source, options exist for deploying both lines simultaneously. Departments that routinely utilize this tactic may elect to configure their pumpers with a split bed of LDH, which would allow the engine to drop both lines at the same time.

Since the majority of departments do not currently operate with split LDH hose beds, another option available is to perform the "drop one, drag one" technique. Although this option effectively allows two LDH lines to be deployed at the same time, firefighters must recognize that it has the potential to injure or kill if it is done incorrectly. This practice requires the secondary or tertiary supply pumper to back up to the bed of one of the engines at the fire scene. The supply pumper will then drop their supply line at this engine. Next, the supply line in the bed of the engine at the fire scene is tied to the supply pumper in some fashion. Once the supply engine reverses out toward the water source, they will effectively be dropping line from their own bed while also dragging line out of the other engine's bed (fig. 9–3).

It is critically important to understand that this technique must only be performed if the water source is in a straight-line direction from the fire scene. If the apparatus dragging the supply line has to make any sharp turns to reach the water source, the operation must be aborted. Failure to do so could result in damage to the supply line or injury to firefighters. In addition, this technique should only be performed if the distance between the scene and the additional water source is relatively short. For optimal safety, one spotter should be assigned to watch the bed of the engine at the scene, and another spotter should walk alongside the supply engine performing the layout. A separate radio channel should be used by the spotters and engine operators involved with the layout. If a coupling gets hung up in either hose bed, or some other issue presents itself, the spotter should radio to the engine laying out to stop immediately. Prior to implementing this

Figure 9–3. A pumper performing a "drop one, drag one" maneuver to deploy a parallel supply line operation

technique on the fireground, it is critically important that members train on it extensively.

The goal of a parallel supply line operation is to deliver a high volume of water to the fire scene using two large diameter supply lines. The intention is for each large diameter supply line to supply one-half the total volume of water to the fire scene. This ultimately reduces the total amount of friction loss in the system, making it easier to deliver these higher flow rates to the attack pumpers. Because of this, parallel supply line operations are quite capable of providing fire scenes with flows greater than 2,000 gpm. In order for these flow rates to be achieved, the operator must ensure that the second supply line is connected and fed in the most hydraulically efficient manner.

Since the initial supply line is likely already connected to the large diameter discharge of the pumper, many operators will use an adapter to connect the second line to a 2½" discharge (fig. 9–4). While connecting this way will technically allow water to be supplied through the line, the total volume able to be delivered to the scene will be limited by the plumbing configuration of the 2½" discharge. As mentioned previously, 2½" discharges must only be capable of flowing a minimum of 250 gpm through their plumbing arrangements. Although 2½" discharges often can supply more than the NFPA minimum discharge rate, 1,000 gpm flows are often not realistic. Since the intent of this operation is to split the total flow rate between the

Figure 9–4. This Pierce Ultimate Configuration (PUC) pumper is performing a parallel supply line operation using its large diameter discharge (bottom right) and one 2½" discharge (top right).

two large diameter supply lines laid, connecting the second line in this fashion does not allow for peak efficiency. Connecting the secondary large diameter supply line directly to a 2½" discharge should be considered a last resort for a parallel supply line operation.

For apparatus equipped with only a single LDH discharge, the preferred option for supplying the second supply line is to use a jumbo siamese or manifold appliance. In this configuration, the siamese or manifold appliance is placed approximately 50' from the supply engine. Depending on the type of appliance used, between two and four medium diameter hose (MDH) lines are run from separate discharges on the pumper to feed directly into the siamese/manifold. The LDH supply line is then connected to the other end of the appliance (fig. 9–5). Essentially, the pump operator has created an additional large diameter discharge using multiple smaller discharges. The advantage of this configuration is that it enables the total flow rate leaving the pumper to be split between multiple discharges on the apparatus. This will help to reduce the total friction loss within the pump itself and will allow the total flow rate supplying the scene to be split between the two supply lines.

Pumpers that are not equipped with any large diameter discharges can use this method when they are forced to pump LDH lines. This situation typically occurs when a department switches to LDH but has not purchased

Figure 9–5. This manifold allows four MDH lines to feed a single 5" line.

a new pumper equipped with a large diameter discharge (fig. 9–6). Some organizations will keep two short lengths of MDH lines preconnected to their passenger's side 2½" discharges, with the other ends connected to a siamese appliance. A pumper tasked with supplying an LDH supply line can simply stretch the siamese toward the LDH line, make the connection, and supply water (fig. 9–7). This will allow for respectable flow rates through an LDH line from a rig that is not equipped with a large diameter discharge. If these apparatus are tasked with performing a parallel supply line operation from a strong hydrant, a second siamese/manifold configuration must be implemented (fig. 9–8).

The biggest disadvantage of using a manifold or siamese for a parallel supply line operation is the fact that multiple hoses must be deployed, and connections must be made. In an effort to streamline the parallel supply line process, some fire departments elect to increase the discharge plumbing on the passenger's side pump panel. For example, the Charlottesville (VA) Fire Department outfits their engines with one 4" plumbed LDH discharge and two 3" plumbed discharges on the passenger's side of the apparatus. This allows the department to easily connect two supply lines to the engine when tasked with performing a parallel supply line operation (fig. 9–9). The primary purpose of this configuration is to be able to supply the

Figure 9–6. This pumper was purchased before the department made the switch to LDH and has no large diameter discharge. A manifold and multiple MDH lines enable this reserve pumper to supply water through the LDH carried on the frontline engines.

Figure 9–7. This reserve pumper is outfitted with a siamese and two short preconnected MDH lengths that enable the operator to rapidly supply the LDH carried by other pumpers.

Figure 9–8. Although this pumper has no large diameter discharge, it is supporting two 5" supply lines using MDH connected to multiple 2½" discharges and two manifolds.

Figure 9–9. This pumper is outfitted with one 4" plumbed large diameter discharge and two 3" discharges. Each discharge can reliably flow 1,000 gpm.

department's elevating platform with 2,000 gpm through a parallel 4" supply line operation (fig. 9–10).

Once the second supply line has been connected to the supply pumper in the best available fashion, the supply pumper must confirm that the fire scene is ready to receive water. The discharge valves feeding the second

Figure 9–10. The plumbing configuration of this pumper makes feeding parallel supply line operations extremely easy.

supply line must be opened slowly after the fire scene calls for water. Opening them too quickly causes a large volume of water to be used to fill the supply line, which could result in the engines at the fire scene losing water momentarily while the supply line is filled. The supply operator can minimize or even avoid this by cracking the discharge valves to slowly fill the second supply line. The filling of the second supply line is complete when the discharge gauges feeding it match the pressure of the master discharge gauge. At this point, the operator can fully open the discharge valves feeding the second supply line and dial in the pump discharge pressure for the operation (fig. 9–11).

Once the operation is in service, the operator should be able to supply larger volumes of water at much lower pump pressures. This is due to the fact that there is a tremendous reduction in friction loss within the entire supply system. It is important for the operator to recognize that even though the pump pressures may be lower, the engine revolutions per minute may still be high. This is due to the fact that discharging large volumes of water, even at lower pump pressures, requires a tremendous amount of horsepower from the apparatus motor. In these instances, having a supply pumper with a larger horsepower motor will become extremely valuable.

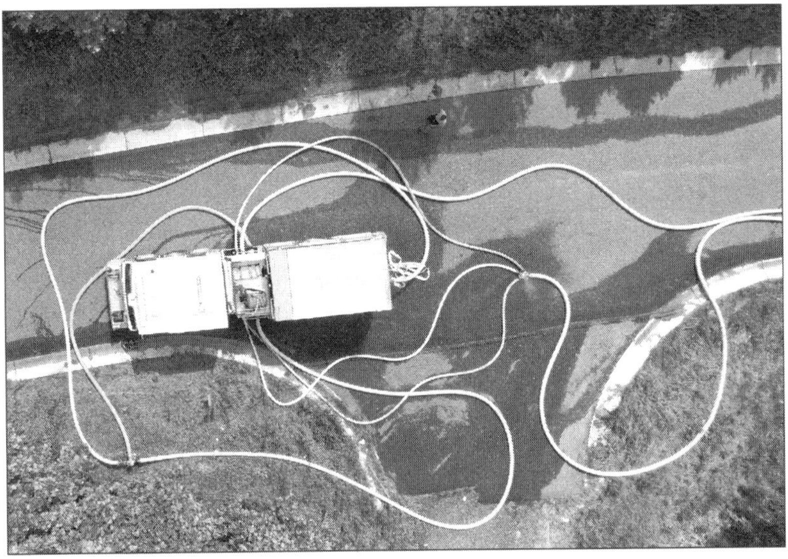

Figure 9–11. An overhead view of a pumper feeding a parallel supply line operation
(*Courtesy*: Kenneth Skipper)

High-Volume Series Pumping Operations

Centrifugal fire pumps can either flow high volumes of water at low pump discharge pressures or flow low volumes of water at high pump discharge pressures. They are not efficient at flowing high volumes at high pressures. While this is true during all pump operations, it has the biggest impact during a drafting operation. This is due to the fact that the apparatus not only has to build pressure on the discharge side of the pump, but it also must create a vacuum on the intake side of the pump.

Scenarios may occur in which the supply operation must provide large volumes of water at high pump discharge pressures from a static source. Examples of this include large-volume relay operations from a static source or incidents that require heavy-caliber streams located by a nearby static source. In these situations, we know that a centrifugal fire pump is capable of delivering its full rated capacity at a pump discharge pressure of 150 psi. Depending on the hose layout and volume of water needed at the fire scene, 150 psi may not be enough to provide the needed fire flow at the scene. If the pump discharge pressure is increased while operating from a draft, however, the flow rate is subsequently diminished.

Conversely, if the pump discharge pressure is reduced below 150 psi, it is possible to flow more than the rated capacity of the pump while drafting. The problem that this creates is that there is not enough pressure available to overcome friction in the hose lines or to produce an effective stream from a large-caliber master stream device. To address this, a *high-volume series pumping operation* can be implemented. This will involve two engines working in tandem to deliver the desired volume at an appropriate pressure. The major difference between these operations and the tandem pumping operations already discussed is that this tactic is intended to flow extremely high volumes of water from a draft.

At least two engines are required to perform a high-volume series pumping evolution, with the first engine being designated as the *drafting engine.* At a minimum, the drafting engine is responsible for performing a twin tube drafting configuration from the static source, but when possible, they should perform a triple tube drafting configuration (fig. 9–12). This will enable the drafting engine to achieve much higher flow rates by minimizing the friction loss in the hard sleeve to the extent possible.[3] All of the water in the system will enter the drafting engine first before being sent to the second pumper in the operation.

The second engine in the operation is referred to as the *boost pumper* and is positioned in close proximity to the drafting engine. Water from

Figure 9–12. This pumper is serving as the drafting engine during a high-volume series pumping operation. Note that the pumper is triple tube drafting and performing a parallel supply line operation.

the drafting engine is supplied at a high volume but low pressure to the intake side of the boost pumper. The boost pumper is then able to add pressure to the volume of water entering its pump in order to provide the necessary nozzle pressure for a heavy-caliber stream or to deliver the water over greater distances.[4] In actuality, the principles of this operation are no different than a supply engine operating from a high-volume, low-pressure fire hydrant. The drafting engine is essentially the high-volume, low-pressure "hydrant," and the boost pumper is the supply engine operating from this "hydrant."

In order for this operation to be successful, it is critically important that the drafting engine perform a parallel supply line operation (fig. 9–13). This will enable the drafting engine to supply these larger flow rates to the boost pumper at much lower pump discharge pressures since the friction loss between the two engines will be tremendously reduced. Because centrifugal fire pumps take advantage of incoming pressure and volume, the boost pumper is capable of adding pressure to the water entering its pump to make the high volume of water being supplied useful on the fireground.[5]

This type of operation, when correctly implemented, can provide tremendous volumes of water to the fire scene, as seen in one particular training evolution in Chesapeake, VA. The Chesapeake Fire Department possesses a trailer-mounted monitor capable of 5,000 gpm as part of its Special Operations Team. This monitor is designed to incorporate up to six large diameter supply lines in order to flow its maximum capacity. During this

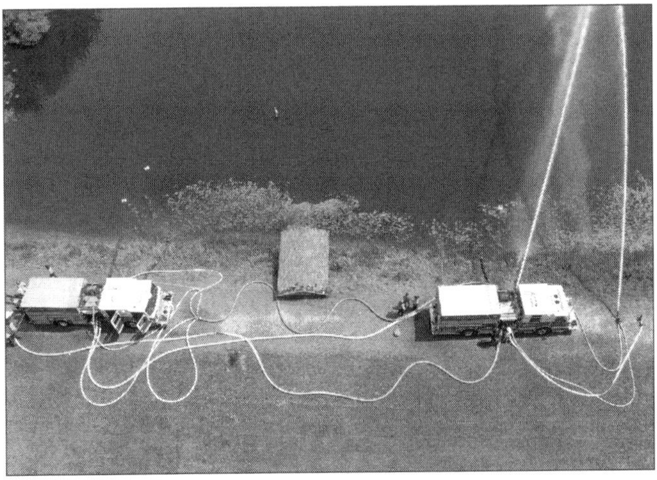

Figure 9–13. An overhead view of a high-volume series pumping operation
(*Courtesy*: Kenneth Skipper)

evolution, four pumpers with pump capacities ranging from 1,250 gpm to 2,000 gpm were utilized. One of the options available to members of the drill was to use all four engines to draft out of the Elizabeth River and supply the monitor. There was not a sufficient amount of hard sleeve available for each of these pumpers to maximize their drafting potential from the source, however. A high-volume series pumping operation was performed instead.[6]

Two 2,000 gpm pumpers were assigned as drafting engines, and each engine triple tube drafted from the river. Both of these pumpers supplied the two boost pumpers via parallel supply lines. Each of the boost pumpers also performed parallel supply line operations from their discharges to the inlets of the trailer-mounted monitor (fig. 9–14). This effectively allowed the high volume of water that was being supplied by the drafting engines to be discharged from the boost pumpers with a minimal amount of internal loss. Operating in this configuration allowed the trailer to reach its maximum 5,000 gpm flow rating (fig. 9–15).[7]

What is important to recognize from the evolution outlined above is that only two of the four engines were actually drafting water from the source. Since the monitor on the trailer was discharging 5,000 gpm, this means that each of the drafting engines was supplying roughly 2,500 gpm, which is 500 gpm above their rated pump capacities. This was only achievable because of the following circumstances:

Figure 9–14. Two high-volume series operations were implemented using this specialty trailer-mounted monitor.
(*Courtesy*: Jillian Stewart)

1. *Each drafting engine performed a triple tube drafting configuration.* This greatly limited the friction loss in the hard sleeve connected to each intake, allowing for the majority of the atmospheric pressure to be used to overcome the lift present.[8]
2. *Each drafting engine performed a parallel supply line operation to the respective boost pumper.* This limited the internal friction loss within the discharge plumbing of the drafting engines to allow more volume to be provided to the boost pumpers.
3. *Each drafting engine had large horsepower motors.* The bigger motors on each of the drafting engines allowed them to provide enough power to spin their impellers fast enough to exceed their pump capacities at low pump discharge pressures.

While this type of operation may have a limited use on the fireground, it is capable of providing substantial flow rates at large-scale incidents. Although most trailer-mounted monitors are intended to be supplied by stand-alone pumping units specifically designed for that purpose, often they arrive on the fire scene at staggered intervals. This means that municipal fire department pumpers will likely be tasked with initially supplying these resources.[9] If operating from a draft and the full capacity of the unit is required, incident commanders should consider implementing this tactic.

Figure 9–15. The high-volume series operations were able to flow 5,000 gpm from the specialty trailer-mounted monitor.
(Courtesy: Jillian Stewart)

Aboveground Looped Systems

When discussing municipal water distribution systems, two basic configurations come to mind: *dead-end systems* and *gridded systems*. Dead-end systems only provide water flow in one direction, while a gridded system provides water from multiple directions. Obviously, the flow rate from a gridded system will exceed that of a dead-end configuration. While this is true regarding the arrangement of water mains in the distribution system, the same principals can be applied to the deployment of supply lines at large-scale fires.[10]

Take a simple water supply operation from a fire hydrant as an example. The most efficient and easiest way for initiating the water supply operation is to place a supply engine at the hydrant to pump water to the attack engine. This configuration creates a unidirectional flow from the water source to the fire scene. The maximum volume of water that is capable of flowing through this setup will be dictated by the flow capacity of the hydrant and how the supply engine connects to the hydrant.

If an additional engine is in need of a water supply at the fire scene, the easiest option is to connect intake-to-intake with the initial attack engine (fig. 9–16). This will allow the second engine at the scene to utilize the unused water from the initial attack engine. Since the flow from the source in this

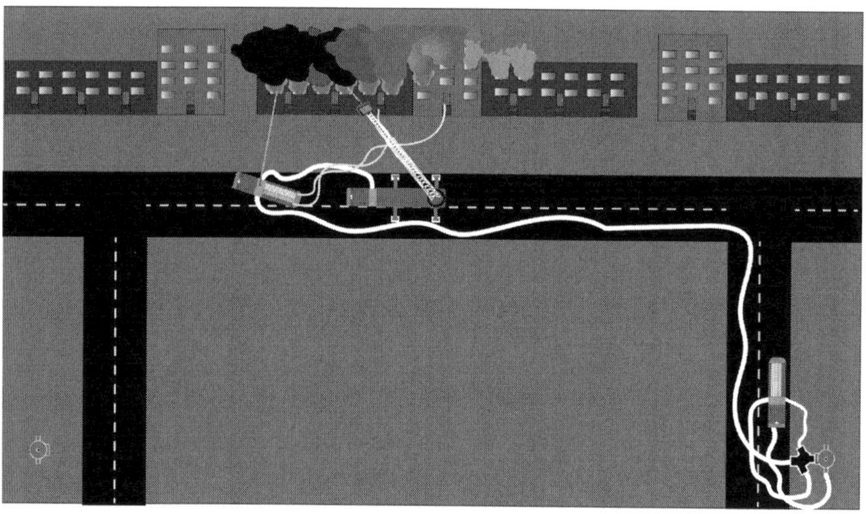

Figure 9–16. A dual pumping operation at the scene allows the supply engine to feed the attack engine and the quint.

scenario is unidirectional, the second attack engine is the last engine to receive water, which essentially creates an aboveground dead-end water system. The initial attack engine will always have first access to the water being supplied to the fire scene since it is upstream from the source.[11]

Building an *aboveground looped system* involves providing the fire scene with water from multiple sources, essentially from multiple directions. In order for this to work, the apparatus at the fire scene must be connected to one another via intake-to-intake connections (fig. 9–17). This allows the water being supplied from each source to be shared between each rig, creating a gridded looped water supply system at the fire scene.[12] This typically is achieved by having a later-arriving engine reverse out to a secondary or tertiary fire hydrant. After reversing out, the second supply pumper should perform a heavy hydrant hookup and supply water to the scene. Ideally, the second supply line is connected to an unused intake on the second attack engine (fig. 9–18). Once water is supplied from the second hydrant, both attack engines at the fire scene should see their residual intake pressures read closer to one another or even equalize completely. This is because water is entering both pumpers from two separate directions. The beauty of this operation is that it is a self-regulating system. Water will flow between the apparatus based on whichever pumper has the higher flow demand (fig. 9–19).[13]

Chapter 7 discussed how main capacity flow tests are performed and explained that high flow rates may exist in sections of the water distribution

Figure 9–17. Apparatus at the scene must be connected intake-to-intake in order for an aboveground looped system to work.

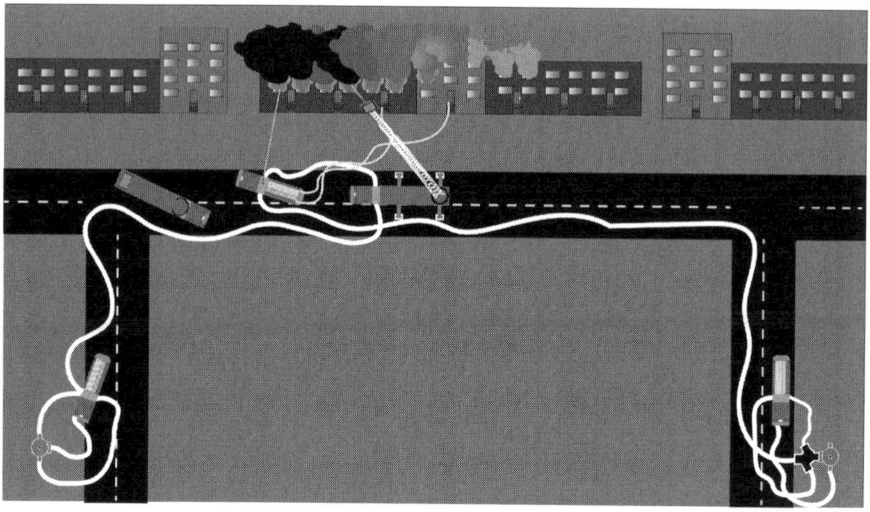

Figure 9–18. The third-arriving engine is used to reverse out to a secondary water source to complete the looped system.

Figure 9–19. An aboveground looped operation is a self-regulating system.

system. Sometimes these flow rates are sufficiently high that they are greater than a single fire hydrant is capable of providing. In order to achieve the full capacity of these strong mains, it is absolutely essential that aboveground looped water supply operations be performed. Furthermore, it is quite likely that the supply pumpers involved in these supply operations will be required to perform parallel supply line operations in order to efficiently deliver the full volume capacity of the water main to the fire scene (fig. 9–20).[14]

Operators must understand that the aboveground looped system is not reserved only for hydrant operations. A drafting operation can successfully support this type of supply configuration in the rural environment, whether the water source is a fire hydrant or a static source. The aboveground looped configuration works because two elements are present:

1. Supply pumpers are present and pumping water from the source to the fire scene.
2. The apparatus at the fire scene are connected to one another via a dual pumping operation.

When utilizing a fire hydrant supplied by a municipal water distribution system, it is not recommended to use a nonpotable static source as the second water supply for an aboveground loop operation. This is because there is the potential for the nonpotable water from the static source to backflow and contaminate the drinking water in the municipal system. For this reason, every attempt should be made to ensure apparatus being supplied from municipal hydrant sources do not intermingle with apparatus being supplied from nontreated static water sources (fig. 9–21). If there is no way to avoid this, and the incident requires an aboveground looped system from both sources, the incident commander must immediately

Figure 9–20. An aboveground looped system viewed from above. Each quint is being supplied through parallel supply line operations from two separate hydrants. The two quints are connected intake-to-intake. A flow rate of 4,000 gpm was achieved through this configuration.
(*Courtesy*: Kenneth Skipper)

contact the water authority to notify them of the situation. The water authority will likely need to issue "boil water" notices to prevent the public from getting sick.

Private water distribution systems can be the exception to this rule. This is because the water in these systems is often untreated and is strictly used for fire protection purposes or sanitation purposes, or a mixture of both. In these instances, if the fire department knows that the water feeding the hydrants is in fact nonpotable, performing an aboveground loop operation in conjunction with a static source is an acceptable option. The key here is knowing without a doubt that the water from both sources is nonpotable.

Operators should remember that it is possible for more than two apparatus to dual pump and be incorporated into the aboveground loop system at the fire scene. The only limiting factor in the operation will be the number of valved intakes equipped on the apparatus at the scene. This is why it is so important for fire departments to specify at least three large intakes, preferably four, on their engines and to ensure that an intake valve is installed on each one. Aboveground loop systems are also capable of incorporating more than two water supplies for the operation. This will again be limited by the number of unused intakes available on the apparatus at the fire scene,

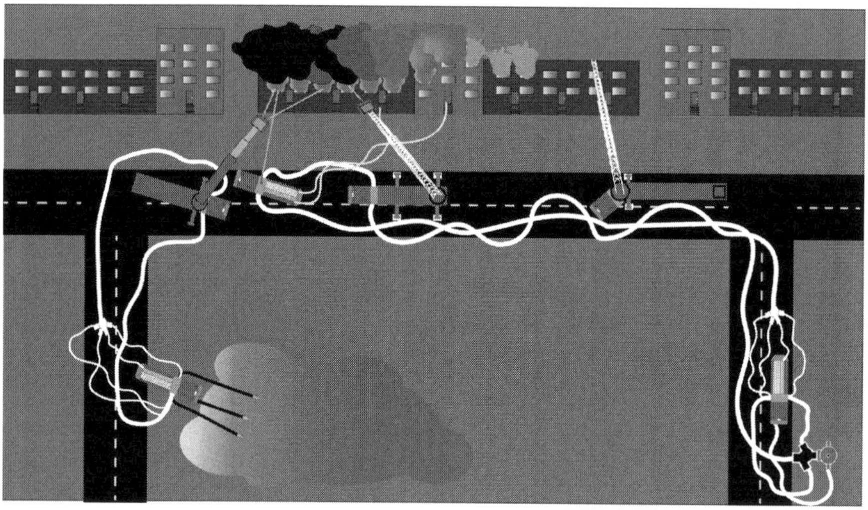

Figure 9–21. An aboveground looped system should be avoided if one of the sources contains nonpotable water. The static source in this example is only supplying water to the tower ladder on the left. Hydrant water is being used to supply the other apparatus operating on the scene.

as well as the number of supply pumpers available to pump the additional water sources.

Supplementing Municipal Water Supplies

Fires in large occupancies typically require extremely high fire flow rates in order to completely control the incident. This will pose a significant challenge for any water supply operation to support. Most firefighters think that if they are operating from a fire hydrant, then there is an unlimited supply of water for these large-scale incidents. In reality, it is very possible for heavy-caliber stream operations of long duration to deplete the water in the municipal water distribution system. This of course is dependent on the capabilities of the distribution system, the flow rate at the scene, and the duration of the flow.

The criteria used by the Insurance Services Office (ISO) to evaluate a jurisdiction's municipal water system can serve as a general reference to determine the time frame that a particular flow rate can be supported. To receive maximum credit in areas requiring needed fire flows of 2,500 gpm or less, the water distribution system should be capable of providing this flow rate for at least two hours. In areas requiring needed fire flows of 3,000 gpm and 3,500 gpm, the municipal water system should be able to provide this flow rate for at least three hours in order to receive maximum credit.[15]

Firefighters should use these time frames as general guidelines to determine how long their water is likely to be reliably available. Given the variations in the water distribution systems of different localities, the time frame may be longer or shorter. In either case, the incident commander must consider what the plan of action will be if the water demands of the operation outpace what the distribution system can provide. Once this point is reached, the operation will need to be supplemented by an additional source, which may be the third or fourth water supply used for the incident. If the distribution system is large enough and zoned appropriately, this could be as simple as performing a relay pumping operation from hydrants located on a separate zone of the distribution system. Other circumstances may require the water supply operation to be supplemented by a static water source. Depending on the resources available to the locality, this may be an extremely large undertaking to successfully achieve.

Regardless of where the supplemental source is coming from, incident commanders must recognize that deploying these operations will take a considerable amount of time and resources. For this reason, they must be initiated well before the municipal system is depleted. As a general recommendation, if heavy-caliber master streams are consistently in service for one hour, the incident commander and water supply officer should be thinking about a plan for supplementing the water supply operation. This will give the water supply officer sufficient time to determine the best source for the supplemental water, as well as give the incident commander time to mobilize the appropriate resources.

The goal of incorporating a supplemental water source is not necessarily to provide the total needed fire flow from a completely different water source. Rather, these operations are intended to help reduce the consumption demand on the municipal system and hopefully give it an opportunity to replenish. Each of the following sections highlights water supply options available for incorporating a supplemental water source during a large-scale incident (fig. 9–22).

Supplemental relay operations

Relaying water from a completely separate source is one of the simplest ways to increase and supplement the available water to a large-scale fire scene. Some jurisdictions have distribution networks that supply water to

Figure 9–22. Large-scale fires may require water from supplemental sources.

customers from several different providers. Each provider has its own water storage and means for distributing water throughout its portion of the distribution system. The ideal option for supplementing a water supply operation would be to pump water from hydrants that are fed by a separate provider.

For example, imagine a manufacturing facility that has its own private water distribution system on-site for fire protection purposes. The fire hydrants outside the facility are fed by the local city water authority, making the two systems completely separate. For a fire within the facility, the private water system will likely be the initial water supply for the operation. If the initial water supply for the incident needed to be supplemented, a relay pumping operation could be implemented to supply water from a hydrant located outside the facility and fed by the city's distribution system. If done appropriately, this will help prevent the initial system from being completely depleted. This tactic will only be successful if the two systems are separate from one another.

Another option is to utilize a nearby static water source for a supplemental water supply. This ensures that the water from the supplemental source is indeed separate from the initial water supply. This is only a viable option if the fire scene is within a realistic distance from a static source that is large enough to supplement the fire flow, however. In addition, it will require each of the supply pumpers assigned to this operation to have the appropriate drafting equipment available. Fire departments must preplan these sources prior to large incidents to determine if this is a viable option.

Supplemental relay operations may require parallel supply line operations to be performed regardless of whether the supplemental supply is a static or separate municipal source. This will be determined based on the needed flow, the amount that must be supplemented at the scene, and the flow capability of the source. If an aboveground looped system is implemented, efforts should be made to prevent potable and nonpotable sources from mixing on the intake side of the attack pumpers.

Supplemental water shuttle operations

Relay pumping from a separate distribution system may not be a realistic option in some localities. In addition, the nearest suitable static water source may be too far from the incident to realistically perform a relay operation. In these situations, a rural water shuttle operation should be implemented to supplement the water available for the incident. The principles for deploying this type of operation are the same as discussed in chapters 4 through 6. The only difference is the setting in which the operation is taking place.

Implementing a water shuttle operation to supplement a water supply in the urban environment requires a tremendous number of resources. Depending on the supplemental flow rate required, several pumpers will be needed at the fill and dump sites, as well as a plethora of mobile water shuttle apparatus. A chart such as table 9–2 can be utilized to determine the number of tankers needed to supplement a certain flow rate based on the round-trip distance and tanker capacity. Since it will likely take some time for all these resources to mobilize, respond, and then deploy, the decision to implement a supplemental water shuttle operation must be made early in an escalating incident.

Table 9–2. Needed number of tankers to provide various supplemental flow rates based on round-trip distance

Tanker capacity (gal)	Supplemental flow rate (gpm)	1 Mile	2 Miles	3 Miles	4 Miles	5 Miles	10 Miles
1,500	500	2	3	4	4	5	8
	1,000	5	6	7	8	10	16
	2,000	9	12	14	17	19	32
2,000	500	2	3	3	3	4	6
	1,000	4	5	6	7	8	13
	2,000	8	10	12	14	16	25
2,500	500	2	2	3	3	3	5
	1,000	4	4	5	6	7	11
	2,000	7	9	10	12	13	21
3,000	500	2	2	2	3	3	5
	1,000	3	4	5	5	6	9
	2,000	7	8	9	11	12	18
3,500	500	2	2	2	2	3	4
	1,000	3	4	4	5	5	8
	2,000	7	8	9	10	11	16
4,000	500	2	2	2	2	3	4
	1,000	3	4	4	5	5	7
	2,000	6	7	8	9	10	15

Note: Assumes that mobile water shuttle apparatus are being filled at the NFPA-recommended 1,000 gpm fill rate.

A lack of interoperability tends to be the biggest hinderance to the implementation of this tactic at large-scale fire scenes, especially for urban fire departments that do not routinely draft. This is primarily due to the fact that water shuttle operations are a foreign concept to most of these firefighters. In addition, many urban fire departments may not carry all the necessary equipment to effectively serve as part of these operations. It is also important to recognize that resources for a water shuttle operation will likely be provided from outside agencies. These organizations may or may not be familiar with operating alongside the fire department that has jurisdiction over the incident. Fire departments that could potentially use this supplemental water supply tactic should explore conducting periodic water shuttle operations with their mutual aid partners. This will ensure that all parties involved are proficient with these operations and could help prevent issues from arising on these once-in-a-career fire scenes.

Some areas of the United States utilize this tactic frequently enough to have developed task forces and strike teams specifically for these operations. These tactical groups are extremely well versed in rural water movement operations and are capable of performing them at peak efficiency during high-profile fires. The foundation for the success of these groups is their dedication to recurrent training on water shuttle operations.

Fireboats

While fireboats are not an option available to every fire department, they are extremely useful tools during supplemental water supply operations. Most firefighters think that fireboats are only useful when fighting fires on the water or in areas immediately surrounded by water. Although these units are useful for mounting marine-based fire attacks, these operations only account for a small percentage of their actual use. For the majority of firefighting applications, fireboats are most advantageous for supplying water to land-based fire apparatus.

We stated earlier that relay pumping water from a static water source is one of the best options for performing a supplemental water supply operation. In some cases, however, the supply pumper tasked with this assignment may not be able to efficiently reach the water source or might not have the necessary drafting equipment. If a fireboat is available, it can be used to pump water out of the source directly to an engine located on land (fig. 9–23). This pumper is then capable of supplying water to the fire scene via a relay pumping operation. From the standpoint of the supply pumper,

Figure 9–23. Fireboats are great options for supporting land-based pumping apparatus.

this would be no different than receiving a pressurized water supply from a fire hydrant. Fireboats can also be used at the fill site for supporting supplemental water shuttle operations. Depending on the performance characteristics of the fireboat's onboard pump, this unit could potentially take the place of the fill site pumper entirely.

Fireboats are not usually resources that can be deployed quickly. This makes it extremely important for these units to be called for early if they are expected to be utilized successfully. Two options exist for using a fireboat if a supplemental water supply operation is already in service prior to the unit's arrival. The first option is that the fireboat can take the place of one of the engines in the supply operation. For long duration fires with multiple operational periods, this could allow municipal engine companies to return to service and be available for other incidents. The other option available is to incorporate the fireboat into the supply operation to supplement the water supply even further (fig. 9–24). This decision should be made based on the needs of the incident.

It is critically important for firefighters to be familiar with the operating capabilities of the fireboat's onboard pump. Smaller boats may have nothing more than a 500 gpm skid unit installed. Conversely, larger fireboats may be equipped with pumps that are capable of exceeding the capacities of municipal fire engines. It is also important to know the pump pressure associated with the fireboat's pump rating, as these pump pressures can differ from

Figure 9–24. A pumper performing a traditional draft and a fireboat are being used to supply an additional engine.

municipal fire apparatus. Pumps with high flow ratings but low discharge pressures may require that a high-volume series operation be performed in order to supply higher volumes of water to the fire scene.

Hydraulically driven submersible pumping units

While not widely used by most fire departments, some organizations have access to hydraulically driven submersible pumps. These units are primarily used for accessing water from static sources and consist of three major components: the *drive motor*, the *hydraulic pumping unit*, and the *submersible centrifugal water pump*. The drive motor is typically diesel-powered and is utilized to power the hydraulic pumping unit. Hydraulic hoses connect the hydraulic power unit to the submersible water pump. Prior to placing the submersible pump into the water, extremely large diameter supply lines are connected to the discharge port of the pump. Once the submersible pump is placed into the water source, hydraulic fluid is sent from the hydraulic pump to spin the impeller within the submersible water pump (fig. 9–25). Water from the submersible unit is then discharged through the large diameter supply line.

These units come in a variety of configurations and capacities. Some of these units can be placed in the bed of a pickup truck to serve as a single-purpose unit. Larger models are mounted on flatbed trailers to offer

Figure 9–25. A hydraulically-driven submersible pump in operation

higher flow rates and more versatility. Hydraulically-driven submersible pumping units in these configurations can reliably provide flows up to 2,500 gpm from static sources. These flows are achieved at relatively low pump discharge pressures. When used for supplemental water supply operations, these units are most effective for pumping their volume into a municipal pumper on land, which creates a high-volume series pumping operation (fig. 9–26). One of the major advantages that a submersible pumping unit provides is that the municipal engine company does not have to perform a drafting operation.

Even larger units are constructed and designed to provide extremely high flow rates for firefighting operations. These units are built on special-purpose trailers or roll-off containers and typically consist of multiple diesel motors that power two separate pumping units. One of the diesel motors is used to power the submersible pump, which sends its water to a boost pump located on the trailer unit. The boost pump is powered by the other diesel motor and is used to add pressure to the water provided by the submersible pump. These larger units essentially perform a high-volume series pumping operation that enables them to flow large volumes of water at considerable pump pressures. The Chesapeake (VA) Fire Department possesses one of these units that is capable of flowing 5,000 gpm using this technology (fig. 9–27).

Hydraulically driven submersible pumping units offer a wide range of benefits to fire departments that invest in them. Nonetheless, their use is

Figure 9–26. A hydraulically-driven submersible pump is being used to supply a pumper and a quint with 2,500 gpm.

Figure 9–27. The Chesapeake (VA) Fire Department's 5,000 gpm hydraulically-driven submersible pumping unit. As shown here, this unit is being used in conjunction with six other municipal pumpers and two smaller submersible units to deliver 15,000 gpm.
(*Courtesy*: Jillian Stewart)

not currently widespread in the U.S. fire service. Units that are slightly more prevalent are trailer-mounted centrifugal fire pumps driven by an auxiliary engine. The operating principles of these pumps are the same as municipal fire engines, with the only difference being that they are intended to be left pumping at a water source for long durations. This prevents a municipal pumper from being tied up on a lengthy incident.

One disadvantage of trailer-mounted centrifugal pumps is that they need to be relatively close to the water source in order to draft the water, similar to a fire department pumper. Since hydraulically driven submersible pumps power the submersible unit through hydraulic hoses, the actual unit containing the diesel motor and hydraulic pump can be anywhere from 100' to 150' from the water source. This allows these units an unparalleled advantage regarding water access during a supplemental water supply operation.

Regardless of whether the centrifugal pumps used are hydraulically-driven submersible pumps or pumps powered by an auxiliary engine, it must be understood that they will also take time to deploy. This means that the supplemental water supply operation may have to be initiated with traditional fire department pumpers and transitioned to these units once they arrive.[16] When used correctly, these pieces of equipment can provide a tremendous amount of supplemental water for an operation and can provide it for a long time.

Water supply support units

The limiting factor regarding supplemental water supply operations tends to be the availability of equipment. This is especially true if the operation requires urban engine companies to perform any sort of drafting operation. In order to combat this, some organizations utilize special water supply support units to assist with supplemental water supply operations. The abilities of these units vary based on the needs of the department, and the units themselves can take many different forms. Some fire departments will use retired or reserve engine companies to serve in this role, while others use trailers loaded with water supply equipment.

The major advantage of these units is that they provide the necessary equipment for large-scale water supply operations without impeding the functionality of day-to-day engine company tasks. It is also important to recognize that these units are useful in both rural and urban fire departments. Fire departments that invest in these resources must ensure that they are appropriately outfitted with equipment, regardless of the setting in which they will be utilized.

Considerations for Prolonged Pump Operations

Water supply operations that last for several operational periods can create logistical problems for operators pumping the equipment and for chief officers managing the scene. Mitigating these problems requires pump operators to be mindful of the status of their apparatus and to communicate their needs to the water supply officer. In addition, incident commanders should preemptively forecast what resources will be needed in order to continue the water supply operation. This requires pump operators, company officers, and chief officers to recognize potential issues and strategize solutions prior to an actual incident. Doing so will allow even the most complex water supply operations to run smoothly. Variables that must be considered include the fuel supply, overheating of the apparatus, the effects of exhaust heat, wrecker service requirements, water runoff from the fire building, and the effects of freezing temperatures.

Specific challenges of long duration incidents

Fuel supply. Incidents of long duration will naturally lead to high fuel consumption by the apparatus on scene. Supply pumpers may be operating at quite high engine revolutions per minute in order to deliver the necessary volume of water to the scene, which increases their fuel consumption rate. Pump operators must diligently check their fuel status throughout the operation to ensure that no issues arise. The water supply officer should also be keeping track of the fuel levels of all the apparatus assigned to the supply operation. Prior to large-scale incidents, fire departments should identify what their contingency plan is for refueling apparatus during an incident. Some organizations have contracts with local fuel depots, while others may have their city public works department bring fuel to the scene. Yet other organizations have their own fire department fuel truck available for refueling operations. Regardless, a refueling plan should be in place prior to an incident to ensure this is accomplished successfully. Along the same lines as fuel, pump operators operating newer apparatus must also pay attention to the amount of diesel exhaust fluid (DEF) used during a prolonged operation. If this is allowed to get too low, damage to the apparatus may result, and the apparatus could cease functioning.

Apparatus overheating. When supply pumpers operate for long durations at high engine revolutions per minute, there is the potential for the apparatus motor to overheat, especially on sweltering summer days. Along with checking the fuel status, the operator should also check the coolant temperature. If the motor is allowed to overheat, considerable damage could result to the apparatus, and there could be an interruption of water to the fire scene as well. If the coolant temperature begins to rise and the radiator fan is unable to cool the motor on its own, there are a couple of ways the operator can assist the cooling process. The simplest option would be to reduce the engine revolutions per minute so that the motor is not working as hard. This may not be a realistic option, however, depending on the situation.

Another option is to position a positive pressure fan in front of the apparatus grill to help blow cool air toward the radiator (fig. 9–28). A third option is to position a booster line with a narrow fog pattern toward the grill of the apparatus. Water from the nozzle should be gently directed toward the grill to help cool the motor (fig. 9–29). The nozzle *does not* have to flow a large volume of water. In addition, *it is critically important that no water be directed anywhere near the air intake of the motor.* Severe motor damage may result. Before performing this cooling method, confirm where the air intake of the apparatus is located. To help prevent the possibility of damage

Figure 9–28. A positive pressure fan is being used to help prevent this pumper from overheating.

Figure 9–29. A narrow fog stream is being used to help prevent the apparatus from overheating.

occurring to any apparatus involved in a prolonged operation, the incident commander should consider having a fire department mechanic placed on standby or brought to the scene.

Exhaust heat. Pump operators should be mindful of where their exhaust is pointed and what surrounds it. Long duration incidents, especially at high engine revolutions per minute, will cause hot exhaust gas. This could cause nearby brush to burn, pavement to melt, or nearby hose lines to be damaged.[17] The best way to combat this is to place a fog nozzle on an unused discharge and spray it toward the exhaust pipe. This can be used to cool the surrounding area and prevent any damage or brush fires from occurring. The flow rate from the nozzle only needs to be enough to keep the area wet.

Wrecker service. While the intention is never to get an apparatus stuck, it does happen from time to time. These situations can be stressful enough, and the last thing the incident commander needs is to take time out to figure out which wrecker service to call. Just as contingency plans for fuel should be made, a plan for a wrecker service should be established prior to any incident. Not only is this beneficial from a logistics standpoint, but it will help the wrecker service become familiar with the fire apparatus used by the department. Apparatus manufacturers may have specific towing considerations based on the options found on the apparatus. Knowledge of this will ensure that no damage is done to the apparatus should a wrecker be needed.

Water runoff from the fire building. Large-scale water flow operations should produce a significant amount of runoff as water leaves the building. A lack of water leaving the fire building indicates that the structure is retaining that water, which is a huge contributing factor to the possibility of a building collapse. Everyone on the fireground must be mindful of this as heavy-caliber streams are operated for prolonged periods of time. Even so, a sufficient amount of water leaving the building could cause other problems at the scene.

During freezing weather, water runoff will obviously cause icing problems. Another concern is that water could start pooling around apparatus, and this is especially dangerous around aerial apparatus. Standing water that is allowed to submerge the outriggers of an aerial apparatus has the potential to undermine the footing of the aerial apparatus, potentially causing a catastrophic collapse. This typically happens when supply lines feeding the aerial device block the flow of water leaving the fire building. A simple solution is to prop up the hose lines in some fashion to allow water to flow under them.

Cold weather. With regard to water supply operations, freezing weather has the potential to cause severe issues. In the urban environment, operators may have to deal with frozen fire hydrants. In the rural environment, water spillage at both the dump site and fill site will cause hazardous ground conditions for firefighters. Apparatus operating in the cold are subject to a myriad of mechanical failures, regardless of the operating setting. Wintry weather operations may require that salt or sand trucks be dispatched to help prevent the roads from icing over. This will be critically important during a shuttle operation. In addition, all of the pumping apparatus involved in the operation must ensure that they constantly flow water in some fashion to prevent freezing in the pump or hose lines. Finally, apparatus may be able to get into position at the beginning of the operation, but freezing conditions could make their exit difficult or impossible. For this reason, it may be beneficial to contact a wrecker company and place them on standby early in the incident, if possible.

Study Questions

1. According to *NFPA 1900*, discharges plumbed with 4" pipe must be capable of flowing at least ________.

   a. 250 gpm
   b. 500 gpm
   c. 625 gpm
   d. 1,000 gpm

2. Generally speaking, due to the plumbing of most large diameter discharges, what is the maximum reliable flow rate that a single LDH supply line is capable of providing?

   a. 500 gpm
   b. 1,000 gpm
   c. 2,000 gpm
   d. The rated capacity of the pump

3. For apparatus with a single large diameter discharge, which of the following is the best option for implementing a parallel supply line operation?

   a. Connect the second LDH directly to a 2½" discharge using an adapter.
   b. Perform a dual pumping operation.
   c. Use multiple sections of MDH from multiple 2½" discharges to feed a siamese or manifold appliance, which then feeds the second LDH.
   d. A parallel supply line operation cannot be done with this pumper.

4. Which of the following tactics is implemented as part of a high-volume series pumping operation?

   a. Parallel supply line operations
   b. Multiple intake drafting
   c. Relay pumping
   d. All of the above

5. Engines at the fire scene are required to perform __________ operations in order for an aboveground looped system to operate successfully.

    a. Dual pumping
    b. Tandem pumping
    c. Relay pumping
    d. Parallel supply line

6. Above ground looped systems should be avoided if one source is ________ and the other source is _________.

    a. A blue top hydrant, a green top hydrant
    b. A blue top hydrant, an orange top hydrant
    c. Potable water, nonpotable water
    d. Dry hydrant, cistern

7. According to ISO, municipal water distribution systems should be capable of providing needed fire flows in areas requiring 3,000 gpm and 3,500 gpm for at least ______.

    a. 1 hour
    b. 2 hours
    c. 3 hours
    d. 10 hours

8. Which of the following is *not* a valid method for supplementing a municipal water supply?

    a. Supplemental water shuttle operations
    b. Fireboats
    c. Separate fire hydrant fed by the same distribution system
    d. Hydraulically driven submersible pumps

9. Fireboats are most advantageous for:

    a. Launching marine-based fire attacks
    b. Feeding land-based fire apparatus
    c. Their ability to be rapidly deployed
    d. All of the above

10. Hydraulically-driven submersible pumps are most efficient when they are configured as part of a ________________ operation.

    a. High-volume series
    b. Dual pumping
    c. Drafting
    d. Intake-to-intake

A

ANSWERS TO STUDY QUESTIONS

Chapter 1

1. D
2. B
3. C
4. B
5. B
6. D
7. B
8. D
9. A
10. D

Chapter 2

1. B
2. B
3. C
4. D
5. B
6. C
7. D
8. B
9. C
10. D

Chapter 3

1. B
2. C
3. D
4. C
5. C
6. A
7. D
8. B
9. C
10. D

Chapter 4

1. C
2. B
3. A
4. B
5. C
6. D
7. D
8. A
9. C
10. B

Chapter 5

1. C
2. D
3. B
4. C
5. A
6. C
7. A
8. C
9. B
10. C

Chapter 6

1. A
2. C
3. B
4. C
5. C
6. C
7. A
8. B
9. C
10. B

Chapter 7

1. B
2. C
3. B
4. C
5. B
6. D
7. A
8. C
9. B
10. D

Chapter 8

1. C
2. B
3. C
4. C
5. B
6. D
7. A
8. B
9. C
10. D

Chapter 9

1. C
2. B
3. C
4. D
5. A
6. C
7. C
8. C
9. B
10. A

B

SOURCE VOLUMES

Table B–1. Square and rectangular sources (gal) (*continues*)

Width (ft)	Avg. depth (ft)	Length (ft) 10	15	20	25	30	35	40	45	50	55	60	65	70	75	80	85	90	95	100
10	2	1,500	2,250	3,000	3,750	4,500	5,250	6,000	6,750	7,500	8,250	9,000	9,750	10,500	11,250	12,000	12,750	13,500	14,250	15,000
	4	3,000	4,500	6,000	7,500	9,000	10,500	12,000	13,500	15,000	16,500	18,000	19,500	21,000	22,500	24,000	25,500	27,000	28,500	30,000
	6	4,500	6,750	9,000	11,250	13,500	15,750	18,000	20,250	22,500	24,750	27,000	29,250	31,500	33,750	36,000	38,250	40,500	42,750	45,000
	8	6,000	9,000	12,000	15,000	18,000	21,000	24,000	27,000	30,000	33,000	36,000	39,000	42,000	45,000	48,000	51,000	54,000	57,000	60,000
	10	7,500	11,250	15,000	18,750	22,500	26,250	30,000	33,750	37,500	41,250	45,000	48,750	52,500	56,250	60,000	63,750	67,500	71,250	75,000
15	2	2,250	3,375	4,500	5,625	6,750	7,875	9,000	10,125	11,250	12,375	13,500	14,625	15,750	16,875	18,000	19,125	20,250	21,375	22,500
	4	4,500	6,750	9,000	11,250	13,500	15,750	18,000	20,250	22,500	24,750	27,000	29,250	31,500	33,750	36,000	38,250	40,500	42,750	45,000
	6	6,750	10,125	13,500	16,875	20,250	23,625	27,000	30,375	33,750	37,125	40,500	43,875	47,250	50,625	54,000	57,375	60,750	64,125	67,500
	8	9,000	13,500	18,000	22,500	27,000	31,500	36,000	40,500	45,000	49,500	54,000	58,500	63,000	67,500	72,000	76,500	81,000	85,500	90,000
	10	11,250	16,875	22,500	28,125	33,750	39,375	45,000	50,625	56,250	61,875	67,500	73,125	78,750	84,375	90,000	95,625	101,250	106,875	112,500
20	2	3,000	4,500	6,000	7,500	9,000	10,500	12,000	13,500	15,000	16,500	18,000	19,500	21,000	22,500	24,000	25,500	27,000	28,500	30,000
	4	6,000	9,000	12,000	15,000	18,000	21,000	24,000	27,000	30,000	33,000	36,000	39,000	42,000	45,000	48,000	51,000	54,000	57,000	60,000
	6	9,000	13,500	18,000	22,500	27,000	31,500	36,000	40,500	45,000	49,500	54,000	58,500	63,000	67,500	72,000	76,500	81,000	85,500	90,000
	8	12,000	18,000	24,000	30,000	36,000	42,000	48,000	54,000	60,000	66,000	72,000	78,000	84,000	90,000	96,000	102,000	108,000	114,000	120,000
	10	15,000	22,500	30,000	37,500	45,000	52,500	60,000	67,500	75,000	82,500	90,000	97,500	105,000	112,500	120,000	127,500	135,000	142,500	150,000
25	2	3,750	5,625	7,500	9,375	11,250	13,125	15,000	16,875	18,750	20,625	22,500	24,375	26,250	28,125	30,000	31,875	33,750	35,625	37,500
	4	7,500	11,250	15,000	18,750	22,500	26,250	30,000	33,750	37,500	41,250	45,000	48,750	52,500	56,250	60,000	63,750	67,500	71,250	75,000
	6	11,250	16,875	22,500	28,125	33,750	39,375	45,000	50,625	56,250	61,875	67,500	73,125	78,750	84,375	90,000	95,625	101,250	106,875	112,500
	8	15,000	22,500	30,000	37,500	45,000	52,500	60,000	67,500	75,000	82,500	90,000	97,500	105,000	112,500	120,000	127,500	135,000	142,500	150,000
	10	18,750	28,125	37,500	46,875	56,250	65,625	75,000	84,375	93,750	103,125	112,500	121,875	131,250	140,625	150,000	159,375	168,750	178,125	187,500
30	2	4,500	6,750	9,000	11,250	13,500	15,750	18,000	20,250	22,500	24,750	27,000	29,250	31,500	33,750	36,000	38,250	40,500	42,750	45,000
	4	9,000	13,500	18,000	22,500	27,000	31,500	36,000	40,500	45,000	49,500	54,000	58,500	63,000	67,500	72,000	76,500	81,000	85,500	90,000
	6	13,500	20,250	27,000	33,750	40,500	47,250	54,000	60,750	67,500	74,250	81,000	87,750	94,500	101,250	108,000	114,750	121,500	128,250	135,000
	8	18,000	27,000	36,000	45,000	54,000	63,000	72,000	81,000	90,000	99,000	108,000	117,000	126,000	135,000	144,000	153,000	162,000	171,000	180,000
	10	22,500	33,750	45,000	56,250	67,500	78,750	90,000	101,250	112,500	123,750	135,000	146,250	157,500	168,750	180,000	191,250	202,500	213,750	225,000

Width (ft)	Avg. depth (ft)	Length (ft)																		
		10	15	20	25	30	35	40	45	50	55	60	65	70	75	80	85	90	95	100
35	2	5,250	7,875	10,500	13,125	15,750	18,375	21,000	23,625	26,250	28,875	31,500	34,125	36,750	39,375	42,000	44,625	47,250	49,875	52,500
	4	10,500	15,750	21,000	26,250	31,500	36,750	42,000	47,250	52,500	57,750	63,000	68,250	73,500	78,750	84,000	89,250	94,500	99,750	105,000
	6	15,750	23,625	31,500	39,375	47,250	55,125	63,000	70,875	78,750	86,625	94,500	102,375	110,250	118,125	126,000	133,875	141,750	149,625	157,500
	8	21,000	31,500	42,000	52,500	63,000	73,500	84,000	94,500	105,000	115,500	126,000	136,500	147,000	157,500	168,000	178,500	189,000	199,500	210,000
	10	26,250	39,375	52,500	65,625	78,750	91,875	105,000	118,125	131,250	144,375	157,500	170,625	183,750	196,875	210,000	223,125	236,250	249,375	262,500
40	2	6,000	9,000	12,000	15,000	18,000	21,000	24,000	27,000	30,000	33,000	36,000	39,000	42,000	45,000	48,000	51,000	54,000	57,000	60,000
	4	12,000	18,000	24,000	30,000	36,000	42,000	48,000	54,000	60,000	66,000	72,000	78,000	84,000	90,000	96,000	102,000	108,000	114,000	120,000
	6	18,000	27,000	36,000	45,000	54,000	63,000	72,000	81,000	90,000	99,000	108,000	117,000	126,000	135,000	144,000	153,000	162,000	171,000	180,000
	8	24,000	36,000	48,000	60,000	72,000	84,000	96,000	108,000	120,000	132,000	144,000	156,000	168,000	180,000	192,000	204,000	216,000	228,000	240,000
	10	30,000	45,000	60,000	75,000	90,000	105,000	120,000	135,000	150,000	165,000	180,000	195,000	210,000	225,000	240,000	255,000	270,000	285,000	300,000
45	2	6,750	10,125	13,500	16,875	20,250	23,625	27,000	30,375	33,750	37,125	40,500	43,875	47,250	50,625	54,000	57,375	60,750	64,125	67,500
	4	13,500	20,250	27,000	33,750	40,500	47,250	54,000	60,750	67,500	74,250	81,000	87,750	94,500	101,250	108,000	114,750	121,500	128,250	135,000
	6	20,250	30,375	40,500	50,625	60,750	70,875	81,000	91,125	101,250	111,375	121,500	131,625	141,750	151,875	162,000	172,125	182,250	192,375	202,500
	8	27,000	40,500	54,000	67,500	81,000	94,500	108,000	121,500	135,000	148,500	162,000	175,500	189,000	202,500	216,000	229,500	243,000	256,500	270,000
	10	33,750	50,625	67,500	84,375	101,250	118,125	135,000	151,875	168,750	185,625	202,500	219,375	236,250	253,125	270,000	286,875	303,750	320,625	337,500
50	2	7,500	11,250	15,000	18,750	22,500	26,250	30,000	33,750	37,500	41,250	45,000	48,750	52,500	56,250	60,000	63,750	67,500	71,250	75,000
	4	15,000	22,500	30,000	37,500	45,000	52,500	60,000	67,500	75,000	82,500	90,000	97,500	105,000	112,500	120,000	127,500	135,000	142,500	150,000
	6	22,500	33,750	45,000	56,250	67,500	78,750	90,000	101,250	112,500	123,750	135,000	146,250	157,500	168,750	180,000	191,250	202,500	213,750	225,000
	8	30,000	45,000	60,000	75,000	90,000	105,000	120,000	135,000	150,000	165,000	180,000	195,000	210,000	225,000	240,000	255,000	270,000	285,000	300,000
	10	37,500	56,250	75,000	93,750	112,500	131,250	150,000	168,750	187,500	206,250	225,000	243,750	262,500	281,250	300,000	318,750	337,500	356,250	375,000
55	2	8,250	12,375	16,500	20,625	24,750	28,875	33,000	37,125	41,250	45,375	49,500	53,625	57,750	61,875	66,000	70,125	74,250	78,375	82,500
	4	16,500	24,750	33,000	41,250	49,500	57,750	66,000	74,250	82,500	90,750	99,000	107,250	115,500	123,750	132,000	140,250	148,500	156,750	165,000
	6	24,750	37,125	49,500	61,875	74,250	86,625	99,000	111,375	123,750	136,125	148,500	160,875	173,250	185,625	198,000	210,375	222,750	235,125	247,500
	8	33,000	49,500	66,000	82,500	99,000	115,500	132,000	148,500	165,000	181,500	198,000	214,500	231,000	247,500	264,000	280,500	297,000	313,500	330,000
	10	41,250	61,875	82,500	103,125	123,750	144,375	165,000	185,625	206,250	226,875	247,500	268,125	288,750	309,375	330,000	350,625	371,250	391,875	412,500

Table B–1. Square and rectangular sources (gal) *(continued)*

Width (ft)	Avg. depth (ft)	Length (ft) 10	15	20	25	30	35	40	45	50	55	60	65	70	75	80	85	90	95	100
60	2	9,000	13,500	18,000	22,500	27,000	31,500	36,000	40,500	45,000	49,500	54,000	58,500	63,000	67,500	72,000	76,500	81,000	85,500	90,000
	4	18,000	27,000	36,000	45,000	54,000	63,000	72,000	81,000	90,000	99,000	108,000	117,000	126,000	135,000	144,000	153,000	162,000	171,000	180,000
	6	27,000	40,500	54,000	67,500	81,000	94,500	108,000	121,500	135,000	148,500	162,000	175,500	189,000	202,500	216,000	229,500	243,000	256,500	270,000
	8	36,000	54,000	72,000	90,000	108,000	126,000	144,000	162,000	180,000	198,000	216,000	234,000	252,000	270,000	288,000	306,000	324,000	342,000	360,000
	10	45,000	67,500	90,000	112,500	135,000	157,500	180,000	202,500	225,000	247,500	270,000	292,500	315,000	337,500	360,000	382,500	405,000	427,500	450,000
65	2	9,750	14,625	19,500	24,375	29,250	34,125	39,000	43,875	48,750	53,625	58,500	63,375	68,250	73,125	78,000	82,875	87,750	92,625	97,500
	4	19,500	29,250	39,000	48,750	58,500	68,250	78,000	87,750	97,500	107,250	117,000	126,750	136,500	146,250	156,000	165,750	175,500	185,250	195,000
	6	29,250	43,875	58,500	73,125	87,750	102,375	117,000	131,625	146,250	160,875	175,500	190,125	204,750	219,375	234,000	248,625	263,250	277,875	292,500
	8	39,000	58,500	78,000	97,500	117,000	136,500	156,000	175,500	195,000	214,500	234,000	253,500	273,000	292,500	312,000	331,500	351,000	370,500	390,000
	10	48,750	73,125	97,500	121,875	146,250	170,625	195,000	219,375	243,750	268,125	292,500	316,875	341,250	365,625	390,000	414,375	438,750	463,125	487,500
70	2	10,500	15,750	21,000	26,250	31,500	36,750	42,000	47,250	52,500	57,750	63,000	68,250	73,500	78,750	84,000	89,250	94,500	99,750	105,000
	4	21,000	31,500	42,000	52,500	63,000	73,500	84,000	94,500	105,000	115,500	126,000	136,500	147,000	157,500	168,000	178,500	189,000	199,500	210,000
	6	31,500	47,250	63,000	78,750	94,500	110,250	126,000	141,750	157,500	173,250	189,000	204,750	220,500	236,250	252,000	267,750	283,500	299,250	315,000
	8	42,000	63,000	84,000	105,000	126,000	147,000	168,000	189,000	210,000	231,000	252,000	273,000	294,000	315,000	336,000	357,000	378,000	399,000	420,000
	10	52,500	78,750	105,000	131,250	157,500	183,750	210,000	236,250	262,500	288,750	315,000	341,250	367,500	393,750	420,000	446,250	472,500	498,750	525,000
75	2	11,250	16,875	22,500	28,125	33,750	39,375	45,000	50,625	56,250	61,875	67,500	73,125	78,750	84,375	90,000	95,625	101,250	106,875	112,500
	4	22,500	33,750	45,000	56,250	67,500	78,750	90,000	101,250	112,500	123,750	135,000	146,250	157,500	168,750	180,000	191,250	202,500	213,750	225,000
	6	33,750	50,625	67,500	84,375	101,250	118,125	135,000	151,875	168,750	185,625	202,500	219,375	236,250	253,125	270,000	286,875	303,750	320,625	337,500
	8	45,000	67,500	90,000	112,500	135,000	157,500	180,000	202,500	225,000	247,500	270,000	292,500	315,000	337,500	360,000	382,500	405,000	427,500	450,000
	10	56,250	84,375	112,500	140,625	168,750	196,875	225,000	253,125	281,250	309,375	337,500	365,625	393,750	421,875	450,000	478,125	506,250	534,375	562,500

Width (ft)	Avg. depth (ft)	Length (ft) 10	15	20	25	30	35	40	45	50	55	60	65	70	75	80	85	90	95	100
80	2	12,000	18,000	24,000	30,000	36,000	42,000	48,000	54,000	60,000	66,000	72,000	78,000	84,000	90,000	96,000	102,000	108,000	114,000	120,000
	4	24,000	36,000	48,000	60,000	72,000	84,000	96,000	108,000	120,000	132,000	144,000	156,000	168,000	180,000	192,000	204,000	216,000	228,000	240,000
	6	36,000	54,000	72,000	90,000	108,000	126,000	144,000	162,000	180,000	198,000	216,000	234,000	252,000	270,000	288,000	306,000	324,000	342,000	360,000
	8	48,000	72,000	96,000	120,000	144,000	168,000	192,000	216,000	240,000	264,000	288,000	312,000	336,000	360,000	384,000	408,000	432,000	456,000	480,000
	10	60,000	90,000	120,000	150,000	180,000	210,000	240,000	270,000	300,000	330,000	360,000	390,000	420,000	450,000	480,000	510,000	540,000	570,000	600,000
85	2	12,750	19,125	25,500	31,875	38,250	44,625	51,000	57,375	63,750	70,125	76,500	82,875	89,250	95,625	102,000	108,375	114,750	121,125	127,500
	4	25,500	38,250	51,000	63,750	76,500	89,250	102,000	114,750	127,500	140,250	153,000	165,750	178,500	191,250	204,000	216,750	229,500	242,250	255,000
	6	38,250	57,375	76,500	95,625	114,750	133,875	153,000	172,125	191,250	210,375	229,500	248,625	267,750	286,875	306,000	325,125	344,250	363,375	382,500
	8	51,000	76,500	102,000	127,500	153,000	178,500	204,000	229,500	255,000	280,500	306,000	331,500	357,000	382,500	408,000	433,500	459,000	484,500	510,000
	10	63,750	95,625	127,500	159,375	191,250	223,125	255,000	286,875	318,750	350,625	382,500	414,375	446,250	478,125	510,000	541,875	573,750	605,625	637,500
90	2	13,500	20,250	27,000	33,750	40,500	47,250	54,000	60,750	67,500	74,250	81,000	87,750	94,500	101,250	108,000	114,750	121,500	128,250	135,000
	4	27,000	40,500	54,000	67,500	81,000	94,500	108,000	121,500	135,000	148,500	162,000	175,500	189,000	202,500	216,000	229,500	243,000	256,500	270,000
	6	40,500	60,750	81,000	101,250	121,500	141,750	162,000	182,250	202,500	222,750	243,000	263,250	283,500	303,750	324,000	344,250	364,500	384,750	405,000
	8	54,000	81,000	108,000	135,000	162,000	189,000	216,000	243,000	270,000	297,000	324,000	351,000	378,000	405,000	432,000	459,000	486,000	513,000	540,000
	10	67,500	101,250	135,000	168,750	202,500	236,250	270,000	303,750	337,500	371,250	405,000	438,750	472,500	506,250	540,000	573,750	607,500	641,250	675,000
95	2	14,250	21,375	28,500	35,625	42,750	49,875	57,000	64,125	71,250	78,375	85,500	92,625	99,750	106,875	114,000	121,125	128,250	135,375	142,500
	4	28,500	42,750	57,000	71,250	85,500	99,750	114,000	128,250	142,500	156,750	171,000	185,250	199,500	213,750	228,000	242,250	256,500	270,750	285,000
	6	42,750	64,125	85,500	106,875	128,250	149,625	171,000	192,375	213,750	235,125	256,500	277,875	299,250	320,625	342,000	363,375	384,750	406,125	427,500
	8	57,000	85,500	114,000	142,500	171,000	199,500	228,000	256,500	285,000	313,500	342,000	370,500	399,000	427,500	456,000	484,500	513,000	541,500	570,000
	10	71,250	106,875	142,500	178,125	213,750	249,375	285,000	320,625	356,250	391,875	427,500	463,125	498,750	534,375	570,000	605,625	641,250	676,875	712,500
100	2	15,000	22,500	30,000	37,500	45,000	52,500	60,000	67,500	75,000	82,500	90,000	97,500	105,000	112,500	120,000	127,500	135,000	142,500	150,000
	4	30,000	45,000	60,000	75,000	90,000	105,000	120,000	135,000	150,000	165,000	180,000	195,000	210,000	225,000	240,000	255,000	270,000	285,000	300,000
	6	45,000	67,500	90,000	112,500	135,000	157,500	180,000	202,500	225,000	247,500	270,000	292,500	315,000	337,500	360,000	382,500	405,000	427,500	450,000
	8	60,000	90,000	120,000	150,000	180,000	210,000	240,000	270,000	300,000	330,000	360,000	390,000	420,000	450,000	480,000	510,000	540,000	570,000	600,000
	10	75,000	112,500	150,000	187,500	225,000	262,500	300,000	337,500	375,000	412,500	450,000	487,500	525,000	562,500	600,000	637,500	675,000	712,500	750,000

Table B–2. Circular sources (gal)

Source diameter (ft)	Average depth (ft)							
	2	**4**	**6**	**8**	**10**	**12**	**14**	**16**
10	1,178	2,355	3,533	4,710	5,888	7,065	8,243	9,420
20	4,710	9,420	14,130	18,840	23,550	28,260	32,970	37,680
30	10,598	21,195	31,793	42,390	52,988	63,585	74,183	84,780
40	18,840	37,680	56,520	75,360	94,200	113,040	131,880	150,720
50	29,438	58,875	88,313	117,750	147,188	176,625	206,063	235,500
60	42,390	84,780	127,170	169,560	211,950	254,340	296,730	339,120
70	57,698	115,395	173,093	230,790	288,488	346,185	403,883	461,580
80	75,360	150,720	226,080	301,440	376,800	452,160	527,520	602,880
90	95,378	190,755	286,133	381,510	476,888	572,265	667,643	763,020
100	117,750	235,500	353,250	471,000	588,750	706,500	824,250	942,000
110	142,478	284,955	427,433	569,910	712,388	854,865	997,343	1,139,820
120	169,560	339,120	508,680	678,240	847,800	1,017,360	1,186,920	1,356,480
130	198,998	397,995	596,993	795,990	994,988	1,193,985	1,392,983	1,591,980
140	230,790	461,580	692,370	923,160	1,153,950	1,384,740	1,615,530	1,846,320
150	264,938	529,875	794,813	1,059,750	1,324,688	1,589,625	1,854,563	2,119,500
160	301,440	602,880	904,320	1,205,760	1,507,200	1,808,640	2,110,080	2,411,520
170	340,298	680,595	1,020,893	1,361,190	1,701,488	2,041,785	2,382,083	2,722,380
180	381,510	763,020	1,144,530	1,526,040	1,907,550	2,289,060	2,670,570	3,052,080
190	425,078	850,155	1,275,233	1,700,310	2,125,388	2,550,465	2,975,543	3,400,620
200	471,000	942,000	1,413,000	1,884,000	2,355,000	2,826,000	3,297,000	3,768,000

Table B–3. Flowing streams (gpm) (*continues*)

Width (ft)	Avg. depth (ft)	Veclocity (ft/min)						
		5	10	15	20	25	30	35
1	0.5	19	38	56	75	94	113	131
	1	38	75	113	150	188	225	263
	2	75	150	225	300	375	450	525
	3	113	225	338	450	563	675	788
	4	150	300	450	600	750	900	1,050
	5	188	375	563	750	938	1,125	1,313
2	0.5	38	75	113	150	188	225	263
	1	75	150	225	300	375	450	525
	2	150	300	450	600	750	900	1,050
	3	225	450	675	900	1,125	1,350	1,575
	4	300	600	900	1,200	1,500	1,800	2,100
	5	375	750	1,125	1,500	1,875	2,250	2,625
3	0.5	56	113	169	225	281	338	394
	1	113	225	338	450	563	675	788
	2	225	450	675	900	1,125	1,350	1,575
	3	338	675	1,013	1,350	1,688	2,025	2,363
	4	450	900	1,350	1,800	2,250	2,700	3,150
	5	563	1,125	1,688	2,250	2,813	3,375	3,938
4	0.5	75	150	225	300	375	450	525
	1	150	300	450	600	750	900	1,050
	2	300	600	900	1,200	1,500	1,800	2,100
	3	450	900	1,350	1,800	2,250	2,700	3,150
	4	600	1,200	1,800	2,400	3,000	3,600	4,200
	5	750	1,500	2,250	3,000	3,750	4,500	5,250
5	0.5	94	188	281	375	469	563	656
	1	188	375	563	750	938	1,125	1,313
	2	375	750	1,125	1,500	1,875	2,250	2,625
	3	563	1,125	1,688	2,250	2,813	3,375	3,938
	4	750	1,500	2,250	3,000	3,750	4,500	5,250
	5	938	1,875	2,813	3,750	4,688	5,625	6,563
10	0.5	188	375	563	750	938	1,125	1,313
	1	375	750	1,125	1,500	1,875	2,250	2,625
	2	750	1,500	2,250	3,000	3,750	4,500	5,250
	3	1,125	2,250	3,375	4,500	5,625	6,750	7,875
	4	1,500	3,000	4,500	6,000	7,500	9,000	10,500
	5	1,875	3,750	5,625	7,500	9,375	11,250	13,125
15	0.5	281	563	844	1,125	1,406	1,688	1,969
	1	563	1,125	1,688	2,250	2,813	3,375	3,938
	2	1,125	2,250	3,375	4,500	5,625	6,750	7,875
	3	1,688	3,375	5,063	6,750	8,438	10,125	11,813
	4	2,250	4,500	6,750	9,000	11,250	13,500	15,750
	5	2,813	5,625	8,438	11,250	14,063	16,875	19,688
20	0.5	375	750	1,125	1,500	1,875	2,250	2,625
	1	750	1,500	2,250	3,000	3,750	4,500	5,250
	2	1,500	3,000	4,500	6,000	7,500	9,000	10,500
	3	2,250	4,500	6,750	9,000	11,250	13,500	15,750
	4	3,000	6,000	9,000	12,000	15,000	18,000	21,000
	5	3,750	7,500	11,250	15,000	18,750	22,500	26,250

Table B–3. Flowing streams (gpm) *(continues)*

Width (ft)	Avg. Depth (ft)	Velocity (ft/min) 40	45	50	55	60	65	75	80
1	0.5	150	169	188	206	225	244	281	300
	1	300	338	375	413	450	488	563	600
	2	600	675	750	825	900	975	1,125	1,200
	3	900	1,013	1,125	1,238	1,350	1,463	1,688	1,800
	4	1,200	1,350	1,500	1,650	1,800	1,950	2,250	2,400
	5	1,500	1,688	1,875	2,063	2,250	2,438	2,813	3,000
2	0.5	300	338	375	413	450	488	563	600
	1	600	675	750	825	900	975	1,125	1,200
	2	1,200	1,350	1,500	1,650	1,800	1,950	2,250	2,400
	3	1,800	2,025	2,250	2,475	2,700	2,925	3,375	3,600
	4	2,400	2,700	3,000	3,300	3,600	3,900	4,500	4,800
	5	3,000	3,375	3,750	4,125	4,500	4,875	5,625	6,000
3	0.5	450	506	563	619	675	731	844	900
	1	900	1,013	1,125	1,238	1,350	1,463	1,688	1,800
	2	1,800	2,025	2,250	2,475	2,700	2,925	3,375	3,600
	3	2,700	3,038	3,375	3,713	4,050	4,388	5,063	5,400
	4	3,600	4,050	4,500	4,950	5,400	5,850	6,750	7,200
	5	4,500	5,063	5,625	6,188	6,750	7,313	8,438	9,000
4	0.5	600	675	750	825	900	975	1,125	1,200
	1	1,200	1,350	1,500	1,650	1,800	1,950	2,250	2,400
	2	2,400	2,700	3,000	3,300	3,600	3,900	4,500	4,800
	3	3,600	4,050	4,500	4,950	5,400	5,850	6,750	7,200
	4	4,800	5,400	6,000	6,600	7,200	7,800	9,000	9,600
	5	6,000	6,750	7,500	8,250	9,000	9,750	11,250	12,000
5	0.5	750	844	938	1,031	1,125	1,219	1,406	1,500
	1	1,500	1,688	1,875	2,063	2,250	2,438	2,813	3,000
	2	3,000	3,375	3,750	4,125	4,500	4,875	5,625	6,000
	3	4,500	5,063	5,625	6,188	6,750	7,313	8,438	9,000
	4	6,000	6,750	7,500	8,250	9,000	9,750	11,250	12,000
	5	7,500	8,438	9,375	10,313	11,250	12,188	14,063	15,000
10	0.5	1,500	1,688	1,875	2,063	2,250	2,438	2,813	3,000
	1	3,000	3,375	3,750	4,125	4,500	4,875	5,625	6,000
	2	6,000	6,750	7,500	8,250	9,000	9,750	11,250	12,000
	3	9,000	10,125	11,250	12,375	13,500	14,625	16,875	18,000
	4	12,000	13,500	15,000	16,500	18,000	19,500	22,500	24,000
	5	15,000	16,875	18,750	20,625	22,500	24,375	28,125	30,000
15	0.5	2,250	2,531	2,813	3,094	3,375	3,656	4,219	4,500
	1	4,500	5,063	5,625	6,188	6,750	7,313	8,438	9,000
	2	9,000	10,125	11,250	12,375	13,500	14,625	16,875	18,000
	3	13,500	15,188	16,875	18,563	20,250	21,938	25,313	27,000
	4	18,000	20,250	22,500	24,750	27,000	29,250	33,750	36,000
	5	22,500	25,313	28,125	30,938	33,750	36,563	42,188	45,000
20	0.5	3,000	3,375	3,750	4,125	4,500	4,875	5,625	6,000
	1	6,000	6,750	7,500	8,250	9,000	9,750	11,250	12,000
	2	12,000	13,500	15,000	16,500	18,000	19,500	22,500	24,000
	3	18,000	20,250	22,500	24,750	27,000	29,250	33,750	36,000
	4	24,000	27,000	30,000	33,000	36,000	39,000	45,000	48,000
	5	30,000	33,750	37,500	41,250	45,000	48,750	56,250	60,000

Table B–3. Flowing streams (gpm) (*continued*)

Width (ft)	Avg. Depth (ft)	Velocity (ft/min)							
		85	90	95	100	105	110	115	120
1	0.5	319	338	356	375	394	413	431	450
	1	638	675	713	750	788	825	863	900
	2	1,275	1,350	1,425	1,500	1,575	1,650	1,725	1,800
	3	1,913	2,025	2,138	2,250	2,363	2,475	2,588	2,700
	4	2,550	2,700	2,850	3,000	3,150	3,300	3,450	3,600
	5	3,188	3,375	3,563	3,750	3,938	4,125	4,313	4,500
2	0.5	638	675	713	750	788	825	863	900
	1	1,275	1,350	1,425	1,500	1,575	1,650	1,725	1,800
	2	2,550	2,700	2,850	3,000	3,150	3,300	3,450	3,600
	3	3,825	4,050	4,275	4,500	4,725	4,950	5,175	5,400
	4	5,100	5,400	5,700	6,000	6,300	6,600	6,900	7,200
	5	6,375	6,750	7,125	7,500	7,875	8,250	8,625	9,000
3	0.5	956	1,013	1,069	1,125	1,181	1,238	1,294	1,350
	1	1,913	2,025	2,138	2,250	2,363	2,475	2,588	2,700
	2	3,825	4,050	4,275	4,500	4,725	4,950	5,175	5,400
	3	5,738	6,075	6,413	6,750	7,088	7,425	7,763	8,100
	4	7,650	8,100	8,550	9,000	9,450	9,900	10,350	10,800
	5	9,563	10,125	10,688	11,250	11,813	12,375	12,938	13,500
4	0.5	1,275	1,350	1,425	1,500	1,575	1,650	1,725	1,800
	1	2,550	2,700	2,850	3,000	3,150	3,300	3,450	3,600
	2	5,100	5,400	5,700	6,000	6,300	6,600	6,900	7,200
	3	7,650	8,100	8,550	9,000	9,450	9,900	10,350	10,800
	4	10,200	10,800	11,400	12,000	12,600	13,200	13,800	14,400
	5	12,750	13,500	14,250	15,000	15,750	16,500	17,250	18,000
5	0.5	1,594	1,688	1,781	1,875	1,969	2,063	2,156	2,250
	1	3,188	3,375	3,563	3,750	3,938	4,125	4,313	4,500
	2	6,375	6,750	7,125	7,500	7,875	8,250	8,625	9,000
	3	9,563	10,125	10,688	11,250	11,813	12,375	12,938	13,500
	4	12,750	13,500	14,250	15,000	15,750	16,500	17,250	18,000
	5	15,938	16,875	17,813	18,750	19,688	20,625	21,563	22,500
10	0.5	3,188	3,375	3,563	3,750	3,938	4,125	4,313	4,500
	1	6,375	6,750	7,125	7,500	7,875	8,250	8,625	9,000
	2	12,750	13,500	14,250	15,000	15,750	16,500	17,250	18,000
	3	19,125	20,250	21,375	22,500	23,625	24,750	25,875	27,000
	4	25,500	27,000	28,500	30,000	31,500	33,000	34,500	36,000
	5	31,875	33,750	35,625	37,500	39,375	41,250	43,125	45,000
15	0.5	4,781	5,063	5,344	5,625	5,906	6,188	6,469	6,750
	1	9,563	10,125	10,688	11,250	11,813	12,375	12,938	13,500
	2	19,125	20,250	21,375	22,500	23,625	24,750	25,875	27,000
	3	28,688	30,375	32,063	33,750	35,438	37,125	38,813	40,500
	4	38,250	40,500	42,750	45,000	47,250	49,500	51,750	54,000
	5	47,813	50,625	53,438	56,250	59,063	61,875	64,688	67,500
20	0.5	6,375	6,750	7,125	7,500	7,875	8,250	8,625	9,000
	1	12,750	13,500	14,250	15,000	15,750	16,500	17,250	18,000
	2	25,500	27,000	28,500	30,000	31,500	33,000	34,500	36,000
	3	38,250	40,500	42,750	45,000	47,250	49,500	51,750	54,000
	4	51,000	54,000	57,000	60,000	63,000	66,000	69,000	72,000
	5	63,750	67,500	71,250	75,000	78,750	82,500	86,250	90,000

Table B–4. Velocity conversion

Ft/min conversion based on time to travel 10 feet			
Seconds to travel 10 ft	Velocity (ft/min)	Seconds to travel 10 ft	Velocity (ft/min)
1	600	31	19
2	300	32	19
3	200	33	18
4	150	34	18
5	120	35	17
6	100	36	17
7	86	37	16
8	75	38	16
9	67	39	15
10	60	40	15
11	55	41	15
12	50	42	14
13	46	43	14
14	43	44	14
15	40	45	13
16	38	46	13
17	35	47	13
18	33	48	13
19	32	49	12
20	30	50	12
21	29	51	12
22	27	52	12
23	26	53	11
24	25	54	11
25	24	55	11
26	23	56	11
27	22	57	11
28	21	58	10
29	21	59	10
30	20	60	10

NOTES

Chapter 1

1. D. Sylvia, *Fire Service Hydraulics*, 2nd ed. (Tulsa, OK: Fire Engineering Books & Videos, 1970).
2. National Fire Protection Association (NFPA), *NFPA 1900: Standard for Aircraft Rescue and Firefighting Vehicles, Automotive Fire Apparatus, Wildland Fire Apparatus, and Automotive Ambulances, 2024* (Quincy, MA: NFPA, 2023), https://www.nfpa.org/codes-and-standards/all-codes-and-standards/list-of-codes-and-standards/detail?code=1900.
3. NFPA, *NFPA 1900: Standard for Aircraft Rescue and Firefighting Vehicles, Automotive Fire Apparatus, Wildland Fire Apparatus, and Automotive Ambulances, 2024.*
4. C. Brakhage and L.A. Miller, "Fire Pump Theory," in *Pumping and Aerial Apparatus Driver/Operator Handbook*, 3rd ed. (Stillwater, OK: Fire Protection Publications, Oklahoma State University, 2015), 300.
5. Brakhage and Miller, "Fire Pump Theory."
6. NFPA, *NFPA 1900: Standard for Aircraft Rescue and Firefighting Vehicles, Automotive Fire Apparatus, Wildland Fire Apparatus, and Automotive Ambulances, 2024.*
7. NFPA, *NFPA 1900: Standard for Aircraft Rescue and Firefighting Vehicles, Automotive Fire Apparatus, Wildland Fire Apparatus, and Automotive Ambulances, 2024.*
8. NFPA, *NFPA 1900: Standard for Aircraft Rescue and Firefighting Vehicles, Automotive Fire Apparatus, Wildland Fire Apparatus, and Automotive Ambulances, 2024.*
9. Hale Products, Inc., *Operation Installation Maintenance Manual for Hale Master Intake Valve (MIV) 2.0,* (Ocala, FL: Hale Products Inc., July 19, 2021), https://www.haleproducts.com/parts-accessories/miv-2-0#product_tabs_overview.
10. Task Force Tips, *Manual: Ball Intake Valve, Ball Intake Valve RC, Jumbo Ball Intake Valve, Jumbo Ball Intake Valve RC* (Valparaiso, IN: Task Force Tips, 2018).

11. NFPA, *NFPA 1900: Standard for Aircraft Rescue and Firefighting Vehicles, Automotive Fire Apparatus, Wildland Fire Apparatus, and Automotive Ambulances, 2024.*
12. NFPA, *NFPA 1900: Standard for Aircraft Rescue and Firefighting Vehicles, Automotive Fire Apparatus, Wildland Fire Apparatus, and Automotive Ambulances, 2024.*
13. NFPA, *NFPA 1900: Standard for Aircraft Rescue and Firefighting Vehicles, Automotive Fire Apparatus, Wildland Fire Apparatus, and Automotive Ambulances, 2024.*
14. A. Soccodato, "Maximizing Pump Capacity While Drafting: The Twin Tube Setup," *Fire Apparatus & Emergency Equipment* 25, no. 6 (June 1, 2020), https://www.fireapparatusmagazine.com/fire-apparatus/maximizing-pump-capacity-while-drafting-the-twin-tube-setup/.
15. NFPA, *NFPA 1900: Standard for Aircraft Rescue and Firefighting Vehicles, Automotive Fire Apparatus, Wildland Fire Apparatus, and Automotive Ambulances, 2024.*
16. Soccodato, "Maximizing Pump Capacity While Drafting."
17. NFPA, *NFPA 1900: Standard for Aircraft Rescue and Firefighting Vehicles, Automotive Fire Apparatus, Wildland Fire Apparatus, and Automotive Ambulances, 2024.*
18. NFPA, *NFPA 1900: Standard for Aircraft Rescue and Firefighting Vehicles, Automotive Fire Apparatus, Wildland Fire Apparatus, and Automotive Ambulances, 2024.*
19. NFPA, *NFPA 1900: Standard for Aircraft Rescue and Firefighting Vehicles, Automotive Fire Apparatus, Wildland Fire Apparatus, and Automotive Ambulances, 2024.*
20. NFPA, *NFPA 1900: Standard for Aircraft Rescue and Firefighting Vehicles, Automotive Fire Apparatus, Wildland Fire Apparatus, and Automotive Ambulances, 2024.*
21. NFPA, *NFPA 1900: Standard for Aircraft Rescue and Firefighting Vehicles, Automotive Fire Apparatus, Wildland Fire Apparatus, and Automotive Ambulances, 2024.*
22. NFPA, *NFPA 1900: Standard for Aircraft Rescue and Firefighting Vehicles, Automotive Fire Apparatus, Wildland Fire Apparatus, and Automotive Ambulances, 2024.*
23. Fire Research Corp., *Pump Boss Pressure Governor and Engine Monitoring Model: PBA400* (Nesconset, NY: Fire Research Corp.), https://www.fireresearch.com/pdfs/pba400.pdf.
24. Fire Research, *Pump Boss Pressure Governor and Engine Monitoring Model: PBA400.*
25. Fire Research, *Pump Boss Pressure Governor and Engine Monitoring Model: PBA400.*
26. Fire Research, *Pump Boss Pressure Governor and Engine Monitoring Model: PBA400.*

27. NFPA, *NFPA 1900: Standard for Aircraft Rescue and Firefighting Vehicles, Automotive Fire Apparatus, Wildland Fire Apparatus, and Automotive Ambulances, 2024.*

Chaper 2

1. P. Spurgeon, "The Equation," in *Fire Service Hydraulics and Pump Operations* (Tulsa, OK: Fire Engineering Books & Videos, 2012), 32–34.
2. C. Brakhage and L. A. Miller, "Theoretical Pressure Calculations," in *Pumping and Aerial Apparatus Driver/Operator Handbook*, 3rd ed. (Stillwater, OK: Fire Protection Publications, Oklahoma State University, 2015), 222.
3. Brakhage and Miller, *Pumping and Aerial Apparatus Driver/Operator Handbook*, 213.
4. National Fire Protection Association (NFPA), *NFPA 1900: Standard for Aircraft Rescue and Firefighting Vehicles, Automotive Fire Apparatus, Wildland Fire Apparatus, and Automotive Ambulances, 2024* (Quincy, MA: NFPA, 2023), https://www.nfpa.org/codes-and-standards/all-codes-and-standards/list-of-codes-and-standards/detail?code=1900.
5. NFPA, *NFPA 1900: Standard for Aircraft Rescue and Firefighting Vehicles, Automotive Fire Apparatus, Wildland Fire Apparatus, and Automotive Ambulances, 2024.*

Chapter 3

1. W. F. Crapo, "Velocity and Flow," in *Fire Protection Hydraulics and Water Supply*, 3rd ed. (Burlington, MA: Jones and Bartlett Learning, 2017), 57–58.
2. National Fire Protection Association (NFPA), *NFPA 1900: Standard for Aircraft Rescue and Firefighting Vehicles, Automotive Fire Apparatus, Wildland Fire Apparatus, and Automotive Ambulances, 2024* (Quincy, MA: NFPA, 2023), https://www.nfpa.org/codes-and-standards/all-codes-and-standards/list-of-codes-and-standards/detail?code=1900.
3. National Fire Protection Association (NFPA), *NFPA 1961: Standard on Fire Hose, 2020* (Quincy, MA: NFPA, 2019), https://www.nfpa.org/codes-and-standards/all-codes-and-standards/list-of-codes-and-standards/detail?code=1961.
4. National Fire Protection Association (NFPA), *NFPA 1962: Standard for the Care, Use, Inspection, Service Testing, and Replacement of Fire Hose, Couplings, Nozzles, and Fire Hose Appliances, 2018* (Quincy, MA: NFPA, 2017), https://www.nfpa.org/codes-and-standards/all-codes-and-standards/list-of-codes-and-standards/detail?code=1962.
5. NFPA, *NFPA 1962: Standard for the Care, Use, Inspection, Service Testing, and Replacement of Fire Hose, Couplings, Nozzles, and Fire Hose Appliances, 2018.*
6. C. Brakhage and L. A. Miller, "Relay Pumping Operations," ch. 12 in *Pumping and Aerial Apparatus Driver/Operator Handbook*, 3rd ed. (Stillwater, OK: Fire Protection Publications, Oklahoma State University, 2015), 430–431.

7. NFPA, *NFPA 1900: Standard for Aircraft Rescue and Firefighting Vehicles, Automotive Fire Apparatus, Wildland Fire Apparatus, and Automotive Ambulances, 2024.*
8. P. D. Shapiro, "Water Delivery Operations for Fires in High-Rise Structures," in *Large Volume Water Delivery* (Tulsa, OK: Fire Engineering Books & Videos, 2021), 107–108.
9. P. D. Shapiro, "Pumping from the Hydrant," in *Large Volume Water Delivery* (Tulsa, OK: Fire Engineering Books & Videos, 2021), 45–50.
10. A. Soccodato, "Advantages to Suction-to-Suction Pumping," *Virginia Fire News*, 2021.
11. Soccodato, "Advantages to Suction-to-Suction Pumping."
12. Shapiro, "Pumping from the Hydrant."
13. C. Brakhage and L. A. Miller, "Positioning Apparatus," in *Pumping and Aerial Apparatus Driver/Operator Handbook*, 3rd ed. (Stillwater, OK: Fire Protection Publications, Oklahoma State University, 2015), 161.

Chapter 4

1. A. Soccodato, "Maximizing Pump Capacity While Drafting: The Twin Tube Setup," *Fire Apparatus & Emergency Equipment* 25, no. 6 (June 1, 2020).
2. W. F. Crapo, "Theory of Drafting and Pump Testing," in *Fire Protection Hydraulics and Water Supply* (Jones & Bartlett Learning, 2017), 126–137.
3. Crapo, "Theory of Drafting and Pump Testing."
4. National Fire Protection Association (NFPA), *NFPA 1900: Standard for Aircraft Rescue and Firefighting Vehicles, Automotive Fire Apparatus, Wildland Fire Apparatus, and Automotive Ambulances, 2024* (Quincy, MA: NFPA, 2023), https://www.nfpa.org/codes-and-standards/all-codes-and-standards/list-of-codes-and-standards/detail?code=1900.
5. NFPA, *NFPA 1900: Standard for Aircraft Rescue and Firefighting Vehicles, Automotive Fire Apparatus, Wildland Fire Apparatus, and Automotive Ambulances, 2024.*
6. NFPA, *NFPA 1900: Standard for Aircraft Rescue and Firefighting Vehicles, Automotive Fire Apparatus, Wildland Fire Apparatus, and Automotive Ambulances, 2024.*
7. NFPA, *NFPA 1900: Standard for Aircraft Rescue and Firefighting Vehicles, Automotive Fire Apparatus, Wildland Fire Apparatus, and Automotive Ambulances, 2024*; and NFPA, *NFPA 1910: Standard for the Inspections, Maintenance, Refurbishment, Testing, and Retirement of In-Service Emergency Vehicles and Marine Firefighting Vessels, 2024* (Quincy, MA: NFPA, 2023), https://www.nfpa.org/codes-and-standards/all-codes-and-standards/list-of-codes-and-standards/detail?code=1910.
8. D. Sylvia, "Fire Service Pumps," in *Fire Service Hydraulics*, 2nd ed. (Tulsa, OK: Fire Engineering Books & Videos, 1970), 147–150.
9. Sylvia, "Fire Service Pumps."

10. NFPA, *NFPA 1910: Standard for the Inspections, Maintenance, Refurbishment, Testing, and Retirement of In-Service Emergency Vehicles and Marine Firefighting Vessels, 2024.*
11. Soccodato, "Maximizing Pump Capacity While Drafting: The Twin Tube Setup."
12. Crapo, "Theory of Drafting and Pump Testing."
13. Crapo, "Theory of Drafting and Pump Testing."
14. NFPA, *NFPA 1910: Standard for the Inspections, Maintenance, Refurbishment, Testing, and Retirement of In-Service Emergency Vehicles and Marine Firefighting Vessels, 2024.*
15. A. Soccodato, "Burp Drafting: An Alternative Primer-Less Priming Method," *Fire Apparatus & Emergency Equipment* 27, no. 9 (2022).
16. Soccodato, "Maximizing Pump Capacity While Drafting: The Twin Tube Setup."
17. Soccodato, "Burp Drafting: An Alternative Primer-Less Priming Method."
18. Soccodato, "Burp Drafting: An Alternative Primer-Less Priming Method."
19. Soccodato, "Burp Drafting: An Alternative Primer-Less Priming Method."
20. Soccodato, "Burp Drafting: An Alternative Primer-Less Priming Method."
21. Soccodato, "Burp Drafting: An Alternative Primer-Less Priming Method."
22. Soccodato, "Burp Drafting: An Alternative Primer-Less Priming Method."
23. Soccodato, "Burp Drafting: An Alternative Primer-Less Priming Method."
24. Soccodato, "Burp Drafting: An Alternative Primer-Less Priming Method."
25. Soccodato, "Burp Drafting: An Alternative Primer-Less Priming Method."
26. Soccodato, "Burp Drafting: An Alternative Primer-Less Priming Method."
27. Soccodato, "Burp Drafting: An Alternative Primer-Less Priming Method."
28. Soccodato, "Maximizing Pump Capacity While Drafting: The Twin Tube Setup."
29. Soccodato, "Burp Drafting: An Alternative Primer-Less Priming Method."
30. Soccodato, "Maximizing Pump Capacity While Drafting: The Twin Tube Setup."
31. NFPA, *NFPA 1900: Standard for Aircraft Rescue and Firefighting Vehicles, Automotive Fire Apparatus, Wildland Fire Apparatus, and Automotive Ambulances, 2024.*
32. NFPA, *NFPA 1900: Standard for Aircraft Rescue and Firefighting Vehicles, Automotive Fire Apparatus, Wildland Fire Apparatus, and Automotive Ambulances, 2024.*
33. Soccodato, "Maximizing Pump Capacity While Drafting: The Twin Tube Setup."
34. Soccodato, "Maximizing Pump Capacity While Drafting: The Twin Tube Setup."
35. Soccodato, "Maximizing Pump Capacity While Drafting: The Twin Tube Setup."
36. Soccodato, "Maximizing Pump Capacity While Drafting: The Twin Tube Setup."
37. Soccodato, "Maximizing Pump Capacity While Drafting: The Twin Tube Setup."
38. Soccodato, "Maximizing Pump Capacity While Drafting: The Twin Tube Setup."
39. Soccodato, "Maximizing Pump Capacity While Drafting: The Twin Tube Setup."
40. Soccodato, "Maximizing Pump Capacity While Drafting: The Twin Tube Setup."
41. Soccodato, "Maximizing Pump Capacity While Drafting: The Twin Tube Setup."
42. NFPA, *NFPA 1900: Standard for Aircraft Rescue and Firefighting Vehicles, Automotive Fire Apparatus, Wildland Fire Apparatus, and Automotive Ambulances, 2024.*

Chapter 5

1. National Fire Protection Association (NFPA), *NFPA 1900: Standard for Aircraft Rescue and Firefighting Vehicles, Automotive Fire Apparatus, Wildland Fire Apparatus, and Automotive Ambulances, 2024* (Quincy, MA: NFPA, 2023), https://www.nfpa.org/codes-and-standards/all-codes-and-standards/list-of-codes-and-standards/detail?code=1900.

Chapter 6

1. A. Soccodato, "TurboDraft Operations: A Solution to Your Rural Water Access Problems, Part 1," *Fire Apparatus & Emergency Equipment* 27, no. 3 (March 1, 2022), https://www.fireapparatusmagazine.com/features/turbodraft-operations-a-solution-to-your-rural-water-access-problems-part-1/.
2. Soccodato, "TurboDraft Operations, Part 1."
3. Soccodato, "TurboDraft Operations, Part 1."
4. Soccodato, "TurboDraft Operations, Part 1."
5. Soccodato, "TurboDraft Operations, Part 1."
6. Soccodato, "TurboDraft Operations, Part 1."
7. Soccodato, "TurboDraft Operations, Part 1."
8. Soccodato, "TurboDraft Operations, Part 1."
9. Soccodato, "TurboDraft Operations, Part 1."
10. Soccodato, "TurboDraft Operations, Part 1."
11. Soccodato, "TurboDraft Operations, Part 1."
12. Soccodato, "TurboDraft Operations, Part 1."
13. Soccodato, "TurboDraft Operations, Part 1."
14. A. Soccodato, "TurboDraft Operations: A Solution to Your Rural Water Access Problems, Part 2," *Fire Apparatus & Emergency Equipment* 27, no. 4 (April 1, 2022), https://www.fireapparatusmagazine.com/fire-apparatus/turbodraft-operations-a-solution-to-your-rural-water-access-problems-part-2/.
15. Soccodato, "TurboDraft Operations, Part 2."
16. Soccodato, "TurboDraft Operations, Part 2."
17. Soccodato, "TurboDraft Operations, Part 2."
18. Soccodato, "TurboDraft Operations, Part 2."
19. Soccodato, "TurboDraft Operations, Part 2."
20. Soccodato, "TurboDraft Operations, Part 2."
21. C. Brakhage and C. Goodson, "Static Water Supply Sources," in *Pumping Apparatus Driver/Operator Handbook*, 2nd ed. (Stillwater, OK: Fire Protection Publications, Oklahoma State University, 2006), 383.
22. Soccodato, "TurboDraft Operations, Part 2."
23. Soccodato, "TurboDraft Operations, Part 2."
24. National Fire Protection Association (NFPA), *NFPA 1900: Standard for Aircraft Rescue and Firefighting Vehicles, Automotive Fire Apparatus, Wildland Fire Apparatus, and Automotive Ambulances, 2024* (Quincy, MA: NFPA, 2023), https://

www.nfpa.org/codes-and-standards/all-codes-and-standards/list-of-codes-and-standards/detail?code=1900.
25. Soccodato, "TurboDraft Operations, Part 2."
26. W. F. Eckman, "Water Shuttle Operations," in *The Fire Department Water Supply Handbook* (Tulsa, OK: Fire Engineering Books & Videos, 1994), 250–256.
27. Eckman, "Water Shuttle Operations."

Chapter 7

1. M. Christensen, ed., *Installation, Field Testing, and Maintenance of Fire Hydrants: Manual of Water Supply Practices—M17*, 4th ed. (Denver, CO: American Water Works Assn., 2006); and National Fire Protection Association (NFPA), *NFPA 291: Recommended Practice for Water Flow Testing and Marking of Hydrants, 2022* (Quincy, MA: NFPA, 2021), https://www.nfpa.org/codes-and-standards/all-codes-and-standards/list-of-codes-and-standards/detail?code=291.
2. NFPA, *NFPA 291: Recommended Practice for Water Flow Testing and Marking of Hydrants, 2022.*
3. A. Soccodato, "Heavy Hydrant Hookups: Maximizing Hydrant Flows, Part 1," *Fire Apparatus & Emergency Equipment* 27, no. 7 (2022), https://www.fireapparatusmagazine.com/equipment/heavy-hydrant-hookups-maximizing-hydrant-flows-part-1/.
4. Soccodato, "Heavy Hydrant Hookups: Maximizing Hydrant Flows, Part 1."
5. Soccodato, "Heavy Hydrant Hookups: Maximizing Hydrant Flows, Part 1."
6. Insurance Services Office, Inc. (ISO), *Public Protection Classification Summary Report for City of Charlottesville and OPA* (ISO, 2014).
7. Christensen, ed., *Installation, Field Testing, and Maintenance of Fire Hydrants: Manual of Water Supply Practices—M17.*
8. Soccodato, "Heavy Hydrant Hookups: Maximizing Hydrant Flows, Part 1."
9. Soccodato, "Heavy Hydrant Hookups: Maximizing Hydrant Flows, Part 1."
10. Christensen, ed., *Installation, Field Testing, and Maintenance of Fire Hydrants—M17.*
11. Christensen, ed., *Installation, Field Testing, and Maintenance of Fire Hydrants—M17*; and NFPA, *NFPA 291: Recommended Practice for Water Flow Testing and Marking of Hydrants, 2022.*
12. Christensen, ed., *Installation, Field Testing, and Maintenance of Fire Hydrants—M17.*
13. Christensen, ed., *Installation, Field Testing, and Maintenance of Fire Hydrants—M17.*
14. NFPA, *NFPA 291: Recommended Practice for Water Flow Testing and Marking of Hydrants, 2022.*
15. A. Soccodato, "Heavy Hydrant Hookups: Maximizing Hydrant Flows, Part 2," *Fire Apparatus & Emergency Equipment* 27, no. 8 (2022), https://www.fireapparatusmagazine.com/fire-apparatus/heavy-hydrant-hookups-maximizing-hydrant-flows-part-2/.

16. Soccodato, "Heavy Hydrant Hookups: Maximizing Hydrant Flows, Part 2."
17. Soccodato, "Heavy Hydrant Hookups: Maximizing Hydrant Flows, Part 2."
18. Soccodato, "Heavy Hydrant Hookups: Maximizing Hydrant Flows, Part 2."
19. Soccodato, "Heavy Hydrant Hookups: Maximizing Hydrant Flows, Part 2."
20. Soccodato, "Heavy Hydrant Hookups: Maximizing Hydrant Flows, Part 2."
21. Soccodato, "Heavy Hydrant Hookups: Maximizing Hydrant Flows, Part 2."
22. Soccodato, "Heavy Hydrant Hookups: Maximizing Hydrant Flows, Part 2."
23. Soccodato, "Heavy Hydrant Hookups: Maximizing Hydrant Flows, Part 2."
24. Soccodato, "Heavy Hydrant Hookups: Maximizing Hydrant Flows, Part 2."
25. A. Soccodato, "Maximizing Hydrant Flows Using 6-Inch Soft Suction," *Fire Apparatus & Emergency Equipment* 25, no. 10 (2020), https://www.fireapparatusmagazine.com/fire-apparatus/maximizing-hydrant-flows-using-6-inch-soft-suction/.
26. Soccodato, "Maximizing Hydrant Flows Using 6-Inch Soft Suction."
27. Soccodato, "Maximizing Hydrant Flows Using 6-Inch Soft Suction."
28. Soccodato, "Maximizing Hydrant Flows Using 6-Inch Soft Suction."
29. D. Fornell, "Hydrant Hook-ups, Chicago Style," *Firehouse* (January 14, 2013), https://www.firehouse.com/apparatus/type/pumper/blog/10850756/water-supply-and-hydrant-connections.
30. D. Gat, "Backdraft Hydrant Scene," YouTube video, 11 sec., August 4, 2017, https://www.youtube.com/watch?v=uTWvE7TdjuE.
31. National Fire Protection Association (NFPA), *NFPA 1900: Standard for Aircraft Rescue and Firefighting Vehicles, Automotive Fire Apparatus, Wildland Fire Apparatus, and Automotive Ambulances, 2024* (Quincy, MA: NFPA, 2023), https://www.nfpa.org/codes-and-standards/all-codes-and-standards/list-of-codes-and-standards/detail?code=1900.

Chapter 8

1. National Fire Protection Association (NFPA), *NFPA 1900: Standard for Aircraft Rescue and Firefighting Vehicles, Automotive Fire Apparatus, Wildland Fire Apparatus, and Automotive Ambulances, 2024* (Quincy, MA: NFPA, 2023), https://www.nfpa.org/codes-and-standards/all-codes-and-standards/list-of-codes-and-standards/detail?code=1900.
2. NFPA, *NFPA 1900: Standard for Aircraft Rescue and Firefighting Vehicles, Automotive Fire Apparatus, Wildland Fire Apparatus, and Automotive Ambulances, 2024.*
3. NFPA, *NFPA 1900: Standard for Aircraft Rescue and Firefighting Vehicles, Automotive Fire Apparatus, Wildland Fire Apparatus, and Automotive Ambulances, 2024.*
4. NFPA, *NFPA 1900: Standard for Aircraft Rescue and Firefighting Vehicles, Automotive Fire Apparatus, Wildland Fire Apparatus, and Automotive Ambulances, 2024.*

5. NFPA, *NFPA 1900: Standard for Aircraft Rescue and Firefighting Vehicles, Automotive Fire Apparatus, Wildland Fire Apparatus, and Automotive Ambulances, 2024.*
6. NFPA, *NFPA 1900: Standard for Aircraft Rescue and Firefighting Vehicles, Automotive Fire Apparatus, Wildland Fire Apparatus, and Automotive Ambulances, 2024.*
7. C. Brakhage and L. A. Miller, "Operating Aerial Apparatus," in *Pumping and Aerial Apparatus Driver/Operator Handbook*, 3rd ed. (Stillwater, OK: Fire Protection Publications, Oklahoma State University, 2015), 690.
8. D. Sylvia, "Fire Streams," in *Fire Service Hydraulics*, 2nd ed. (Tulsa, OK: Fire Engineering Books & Videos, 1970), 306
9. NFPA, *NFPA 1900: Standard for Aircraft Rescue and Firefighting Vehicles, Automotive Fire Apparatus, Wildland Fire Apparatus, and Automotive Ambulances, 2024.*
10. NFPA, *NFPA 1900: Standard for Aircraft Rescue and Firefighting Vehicles, Automotive Fire Apparatus, Wildland Fire Apparatus, and Automotive Ambulances, 2024.*
11. NFPA, *NFPA 1900: Standard for Aircraft Rescue and Firefighting Vehicles, Automotive Fire Apparatus, Wildland Fire Apparatus, and Automotive Ambulances, 2024.*
12. Task Force Tips, *Manual: Monsoon & Monsoon RC Monitor* (Valparaiso, IN: Task Force Tips, 2016); Elkhart Brass, *8294 Scorpion Losses 4.0" Inlet and 3.5" Outlet* (Elkhart, IN: Elkhart Brass Mfg. Co., 2006); and Akron Brass, *StreamMaster II Electric Fire Monitor System/2000 GPM* (Wooster, OH: Akron Brass Co., 2021), https://www.akronbrass.com/stream-master-ii-electric-2000#product_tabs_download.

Chapter 9

1. National Fire Protection Association (NFPA), *NFPA 1900: Standard for Aircraft Rescue and Firefighting Vehicles, Automotive Fire Apparatus, Wildland Fire Apparatus, and Automotive Ambulances, 2024* (Quincy, MA: NFPA, 2023), https://www.nfpa.org/codes-and-standards/all-codes-and-standards/list-of-codes-and-standards/detail?code=1900.
2. NFPA, *NFPA 1900: Standard for Aircraft Rescue and Firefighting Vehicles, Automotive Fire Apparatus, Wildland Fire Apparatus, and Automotive Ambulances, 2024.*
3. A. Soccodato, "East Coast 15,000 GPM Water Supply Drafting Exercise," *Fire Apparatus & Emergency Equipment* 26, no. 7 (July 2021), https://digital.fireapparatusmagazine.com/fireapparatus/july_2021/MobilePagedReplica.action?pm=2&folio=16#pg16.
4. Soccodato, "East Coast 15,000 GPM Water Supply Drafting Exercise."
5. Soccodato, "East Coast 15,000 GPM Water Supply Drafting Exercise."

6. Soccodato, "East Coast 15,000 GPM Water Supply Drafting Exercise."
7. Soccodato, "East Coast 15,000 GPM Water Supply Drafting Exercise."
8. Soccodato, "East Coast 15,000 GPM Water Supply Drafting Exercise."
9. Soccodato, "East Coast 15,000 GPM Water Supply Drafting Exercise."
10. P. D. Shapiro, "Pumping from the Hydrant," 2 in *Large Volume Water Delivery* (Tulsa, OK: Fire Engineering Books & Videos, 2021), 43–50.
11. Shapiro, "Pumping from the Hydrant."
12. Shapiro, "Pumping from the Hydrant,"
13. A. Soccodato, "Advantages to Suction-to-Suction Pumping," *Virginia Fire News* (2021).
14. Soccodato, "Advantages to Suction-to-Suction Pumping."
15. Insurance Services Office, Inc. (ISO), *Public Protection Classification Summary Report for City of Charlottesville and OPA* (ISO, 2014); and W. F. Eckman, "Fire Hydrants as a Water Source," 6 in *The Fire Department Water Supply Handbook* (Tulsa, OK: Fire Engineering Books & Videos, 1994), 132.
16. Soccodato, "East Coast 15,000 GPM Water Supply Drafting Exercise."
17. A. Soccodato, "TurboDraft Operations: A Solution to Your Rural Water Access Problems, Part 1," *Fire Apparatus & Emergency Equipment* 27, no. 3 (March 1, 2022), https://www.fireapparatusmagazine.com/features/turbodraft-operations-a-solution-to-your-rural-water-access-problems-part-1/.

INDEX